Cracking the SAT II: BIOLOGY Subject Test

THE PRINCETON REVIEW

Cracking the SAT II: BIOLOGY Subject Test

THEODORE SILVER, M.D. AND
THE STAFF OF THE PRINCETON REVIEW

1999–2000 EDITION

RANDOM HOUSE, INC.
NEW YORK

Princeton Review Publishing, L.L.C.
2315 Broadway
New York, NY 10024
E-mail: info@review.com

Copyright © 1999 by Princeton Review Publishing, L.L.C.

All rights reserved under International and Pan-American Copyright Conventions. Published in the United States by Random House, Inc., New York, and simultaneously in Canada by Random House of Canada Limited, Toronto.

ISBN: 0-375-75297-8
ISSN: 1076-5336

SAT II is a registered trademark of the College Board.

Editor: Rachel Warren
Designer: Illeny Maaza
Production Editor: Kristen Azzara
Production Coordinator: Scott Harris
Illustrations by: The Production Department of The Princeton Review

Manufactured in the United States of America on partially recycled paper.

9 8 7 6 5 4 3 2 1

ACKNOWLEDGMENTS

The author thanks John Bergdahl, Bruno Blumenfeld, Cynthia Brantley, Leland Elliott, Alicia Ernst, Effie Hadjiioannou, Julian Ham, Julian Heath, Adam Hurwitz, Peter Jung, Sara Kane, John Katzman, Meher Khambata, Kim Magloire, Russell Murray, John Pak, Christopher Scott, Linda Tarleton, Chris Thomas, Scott Harris, and Illeny Maaza for designing this book.

Also, the editors would like to acknowledge and thank the Research and Development team of Jeannie Yoon, Jane Lacher, and Celeste Ganderson for their assistance in ensuring that our materials and techniques are up-to-date. We would also like to thank Judene Wright for quickly writing the Biology E/M test.

CONTENTS

Chapter 1: Introduction 1
    Point 1: Teaching you Science in the Way the Biology Subject Tests Test It .. 2
    Point 2: Approaching the Questions Strategically 4

Chapter 2: The Questions and Strategies 5
    The Regular SAT II: Biology Subject Test 5
    The Biology E/M Test 9
    Strategy 1: Studying the Right Stuff in the Right Way 10
    Strategy 2: Practice the Right Thing at the Right Time 11
    Strategy 3: Easy Stuff First 11
    Strategy 4: Take a Guess! 11
    Strategy 5: I, II, III—You're Out! 12
    Strategy 6: Avoid the Camouflage Trap 12
    Strategy 7: Avoiding the Temptation Trap 15
    Strategy 8: Make Friends with the Enemy 17
    Strategy 9: Keep Both Eyes Open 19

Chapter 3: Cracking Cellular and Molecular Biology 23
    Organic Chemistry Lesson 1: Amino Acids and Proteins 24
    Organic Chemistry Lesson 2: Carbohydrates 28
    Here's Something You've Seen Before: A Eukaryotic Cell 33
    Where the Cell gets its ATP: Glycolysis, Krebs Cycle, Electron Transport
        Chain, and Oxidative Phosphorylation 45
    Chromosomes and the Whole Organism: The Same Set in Every Cell 57
    How a Whole Cell Reproduces Itself: Mitosis 64
    The Formation of Gametes: Meiosis 69

Chapter 4: Cracking Genetics 75
    Remember why Chromosomes are Important: They Contain Genes 75
    Phenotype and Genes 77
    Punnett Squares 83

Chapter 5: Cracking the Structures and Functions of Organisms 95
    Bodily Support in Animals: Endoskeletons and Exoskeletons 96
    What Nutrition is all About 97
    Nutrition in Plants 103
    Blood and Immunity 105
    What Keeps the Blood Moving? The Heart 109
    The Circulation Summarized 114
    The Nervous System 117
    Moving an Impulse Between Two Neurons: The Synapse 118
    The Central and Peripheral Nervous Systems 122
    Hormones and the Endocrine System 126
    The Female Menstrual Cycle 132
    How Human Beings are Formed: Reproduction and Embryology 136
    What About Reproduction in Plants? 139
    Learning, Behavior, and Coexistence 141
    Behavior in Plants: The Tropisms 143
    Now We'll Talk about Microorganisms: Protists, Fungi, Bacteria,
        and Viruses 146
    Evolution and Ecology 152
    Getting Organized: Phylogeny 154
    Food Chains 160
    Biological Communities Develop: Biomes 163

Chapter 6: The Princeton Review SAT II: Biology Regular Subject Test 165
Chapter 7: Explanations for the SAT II: Biology Regular Subject Test 189
Chapter 8: The Princeton Review SAT II: Biology E/M Subject Test 227
Chapter 9: Explanations for the SAT II: Biology E/M Subject Test 249
Chapter 10: Answers to In-chapter Questions 275

1 Introduction

This book is for students who want to raise their scores on the SAT II: Biology or Biology E/M Subject Test. At The Princeton Review, we know what standardized test makers are up to. That's because we *study* their tests. We take them apart piece by piece and examine them from every angle and perspective: right-side up, upside-down, frontward, backward, and inside-out. *We know how these tests are built*, and that's how we raise scores. Stick with us, and we'll raise yours.

How exactly will we do that? Two ways. We'll:

1. teach you biology *in the way the Biology Subject Tests test it*,

    and

2. show you how to approach the tests *strategically*.

POINT 1: TEACHING YOU SCIENCE IN THE WAY THE BIOLOGY SUBJECT TESTS TEST IT

ETS says its SAT II Biology Subject test covers, among many other topics,

> aerobic respiration, anaerobic respiration, and the biochemical differences between the two.

If you sat with your biology textbook or some other commercial book that promises to prepare you for the Biology Subject Tests, you'd read about a whole lot of stuff *that definitely will not be tested*. You'd read about the mitochondrial matrix and the roles of various enzymes, coenzymes, and cofactors.

You'd see pictures like this:

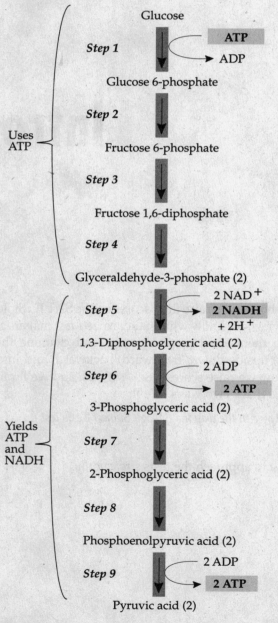

and you'd read text like this:

> Glycolysis is a prime illustration of the manner in which vital biochemical processes occur through a series of steps. The complete catabolism of glucose may be considered to embody nine steps. We'll examine the details of glycolysis and notice that the 6-carbon skeleton of the glucose molecule is sequentially degraded, each step being catalyzed by a specific enzyme, to produce adenosine triphosphate (ATP) via the phosphorylation of adenosine diphosphate. Blah, blah, blah . . . acetyl CoA . . . blah, blah, blah . . . NADH and NAD^+ blah, blah, blah . . . $FADH_2$, blah, blah, blah . . . cytochrome carrier system, blah, blah blah . . . mitochondrial matrix . . . blah, blah, blah . . . inner membrane . . . blah, blah, blah . . . outer membrane . . . blah, blah, blah!

The text would go on and on, scaring and boring you, but offering nothing that would raise your test score. You'd become so sick of it that you'd stop reading (which is fine, because reading that stuff wouldn't raise your test score).

When *we* teach you about aerobic and anaerobic respiration, we tell you exactly what you have to know. As we do that, we drill you (in friendly fashion) to make sure you're with us at every step. When it comes to aerobic and anaerobic respiration, for example, we'll show you that you don't really have to understand anything. You just have to make a couple of simple *associations*, like these:

<u>Ae</u>robic respiration with:
- presence of oxygen, more ATP produced
- Krebs cycle, electron transport chain
- oxidative phosphorylation

<u>An</u>aerobic respiration with:
- absence of oxygen, less ATP produced
- lactic acid

When we get through, you may not really *understand* much about the difference between aerobic and anaerobic respiration. But you don't have to, and we'll prove it. Review the six lines we've just written. Then answer these two Biology Subject Test-like questions:

- Which of the following substances is produced as a result of anaerobic respiration but NOT of aerobic respiration?

    (A) Carbon dioxide
    (B) Lactic acid
    (C) ATP
    (D) Glucose
    (E) Glycogen

- Among the following, which is associated with BOTH aerobic and anaerobic respiration?

    I. ATP production
    II. Krebs cycle
    III. Oxidative phosphorylation

    (A) I only
    (B) II only
    (C) I and II only
    (D) II and III only
    (E) I, II, and III

The answer to the first question is **B**: Anaerobic respiration is associated with the production of lactic acid. The answer to the second question is **A**: Both anaerobic and aerobic respiration produce ATP, but only aerobic respiration is associated with the Krebs cycle and oxidative phosphorylation. Whether or not you *understand* your answers, the scoring machines at ETS will think you *do*. ETS's scoring machines don't look for brilliant scientists and they don't look for understanding. In the case of these two questions they look for a "B" and an "A" in the little oval space on your answer sheet.

POINT 2: APPROACHING THE QUESTIONS *STRATEGICALLY*

It's not enough to study science in the way the Biology Subject Test tests it. You must also study Biology Subject Test questions themselves. You should be wise to their design and familiar with *techniques* that systematically lead to correct answers.

When you sit down to take the Biology Subject Test, you'll see *some* questions whose answers you don't know. In Chapter 1 of this book, we'll show you nine strategies that help you outsmart the Biology Subject Test and its writers. Study our strategies and you may know the right answers even though the question addresses biology that you *don't* know. Our strategies are powerful stuff. They show you how to find right answers logically and systematically, similar to the way in which a good detective solves a crime.

WHAT ABOUT PRACTICE AND PRACTICE TESTS?

This book is interactive. We rehearse you over and over again on the subjects and strategies we teach. We *don't* present a whole long array of drill questions at the end of a chapter. Instead, we watch your progress paragraph by paragraph, page by page. We make sure you're with us every step of the way, and if you're not, we help you figure out *why* you're not. We take what you need to know and drive it into your head, word by word and sentence by sentence. And we do that, incidentally, in a way that makes the whole thing interesting and fun. We then present full-length tests, one Biology test and one Biology E/M test in Chapters 6 and 8. They are complete with explanations that send you back to the appropriate pages of this book in case you need some quick review.

SHOULD I BUY PRACTICE MATERIAL FROM ETS?

It's not a bad idea. If you want to take additional tests beyond the ones we provide, then buy *Official Guide to the SAT II: Subject Tests*, which is published by the College Board. Take the Biology Subject Test and see how easy it is after you've read this book. Or, if you like, take the College Board's test *first*. Score yourself according to College Board's scoring grid. Then go through this book very carefully, and take the test in the back of this book. Score it, and see what a difference this book makes.

OUR JOB AND YOUR JOB

This book is designed to raise your Biology Subject Test score. It's written, it's published, and you're holding it in your hands. That means our job is done. Your job is to read it, tackle the exercises it provides, and learn what it has to teach. We had fun doing our job and, believe it or not, you'll have fun doing yours. So, let the fun begin.

The Questions and Strategies

THE REGULAR SAT II: BIOLOGY SUBJECT TEST

The Regular Biology Subject Test presents 95 multiple-choice questions and gives you one hour to answer them. The test is divided into three sections, A, B, and C, each offering a different question type. It's scored on a scale from 200 to 800 just like the SAT and all of the other Subject Tests that the Educational Testing Service gives.

Parts A, B, and C: The Questions

Part A

Part A of the Biology Subject Test presents you with about 55 ordinary looking multiple-choice questions. You know what these things look like because you've been dealing with them all your life. Still, just for illustration we'll show you two typical A-type biology test questions. You might not be able to answer them right now, but you certainly will be able to answer them after you've finished Chapter 1 of this book.

- Which of the following terms describes the process by which the plasma membrane moves substances inward against a concentration gradient?

    (A) Facilitated transport
    (B) Active transport
    (C) Osmosis
    (D) Diffusion
    (E) Autotrophism

- The endocrine organ that secretes antidiuretic hormone is the

    (A) adrenal cortex
    (B) pancreas
    (C) posterior pituitary gland
    (D) kidney
    (E) liver

As it happens, the correct answers are **B** and **C**, but it doesn't matter right now whether or not you know that. (You *will* know it, *easily*, after you've finished with us.) We just wanted you to know how A-type questions look. Now you know.

Two More Things

1. Some of the A-type questions will be of the "I, II, and III" type, like the following:

    - Which of the following nitrogenous bases are found in DNA?

        I. Thymine
        II. Cytosine
        III. Uracil

        (A) I only
        (B) II only
        (C) I and II only
        (D) I and III only
        (E) I, II, and III

2. Some of the A-type questions will work in "teams" like this:

Questions 45-47:

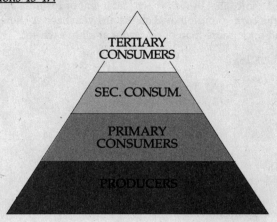

45. Which of the following food chain levels contains the fewest number of organisms?

   (A) blah, blah, blah . . .
   (B) blah, blah, blah . . .
   (C) blah, blah, blah . . .
   (D) blah, blah, blah . . .
   (E) blah, blah, blah . . .

46. Which level in this food chain has the greatest biomass?

   (A) blah, blah, blah . . .
   (B) blah, blah, blah . . .
   (C) blah, blah, blah . . .
   (D) blah, blah, blah . . .
   (E) blah, blah, blah . . .

47. Of the four food chain levels, which of the following contain organisms that are herbivores?

   I. blah
   II. blah
   III. blah

   (A) blah
   (B) blah, blah
   (C) blah, blah
   (D) blah, blah
   (E) blah, blah, blah . . .

OKAY, WHAT ABOUT PART B?

Part B gives you 5 to 6 little "matching tests." Each matching test sets up a list of 5 words or phrases lettered A to E. Then for each list you get 3 to 4 questions, with question numbers next to them. But the questions aren't really questions. They're phrases—half-sentences. Your job is to match the phrase in the question with the word or phrase that appears in the list A to E. Here, we'll forget about science for a second and show you how it works.

Directions: Each set of lettered choices below refers to the numbered statements immediately following it. Select the one lettered choice that best fits each statement or formula and then fill in the corresponding oval on the answer sheet. A choice may be used once, more than once, or not at all in each set.

Questions 1-3

(A) Thyroid
(B) Adrenal cortex
(C) Pancreas
(D) Ovaries
(E) Parathyroid

1. Secretes glucagon

2. Regulates metabolism

3. Structure producing female gametes

The answers, are C, A, and D. Don't worry about the answers. Right now, we just want you to know how B-type questions look. As we said, you'll get about 5 of these B-type matching tests when you take the Biology Subject Test.

AND PART C?

Supposedly, Part C is designed to see if you can think logically about biological experiments. First you're told about an experiment. You might also be shown some graph or table that goes with the experiment. Then you're asked 2 to 4 ordinary looking multiple-choice questions.

Here's an example:

Questions 93-94 refer to an experimenter who wishes to study the process of evolution by working with anaerobic bacterial organisms. He places a small colony of bacteria in Plate 1, which contains a suitable culture medium. After 10 days he removes approximately one-half the bacteria from Plate 1 and places them in Plate 2, which contains both a suitable culture medium and a potent antibiotic related to penicillin.

93. Over the course of time the experimenter should expect to find in both Plates 1 and 2 an increase in the concentration of

(A) ADP
(B) carbon monoxide
(C) lactic acid
(D) Krebs cycle enzymes
(E) oxygen

94. Which of the following time vs. population graphs most likely describes that which the experimenter would observe after initiating his colony in Plate 1?

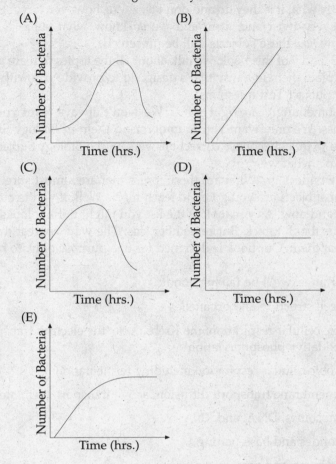

The questions require that you be able to either (1) read a graph and/or (2) exercise a little common-sense logic about the biology you know.

Now that you know what kinds of questions you'll see on the Biology Subject Test, let's talk about the Biology E/M test.

THE BIOLOGY E/M TEST

The Biology E/M (Ecological or Molecular) test is for students who took a biology course that placed emphasis on ecological or molecular biology. If you take this test, you get to choose the area in which you feel more comfortable: Ecology or Molecular Biology.

This test is made up of a common core of 60 questions, which is followed by 20 questions in each of the two specializing sections.

The core section of this test tests the same material as does the regular Biology test. The E/M test is also scored on the same scale as is the regular test—200–800. Now let's talk about the Biology E/M test.

STRATEGY 1: STUDYING THE RIGHT *STUFF* IN THE RIGHT *WAY*

The test writers themselves admit that they don't test all of high school biology. The trouble is they don't tell you exactly what it is they do and don't test. We, however, have studied the Biology Subject Tests. We know every twist and turn, and we *do* know what subjects they will present. More important, we know *how* these subjects will be presented.

In Chapters 2, 3, and 5 of this book, we talk about all the topics that are almost certain to show up on the test. We explain them in a way that's designed to provide you with the test-smarts you need to answer Biology Subject Test questions.

And let's get something clear right now. We won't always want you truly to *understand* the concepts we discuss. We only want you to understand them in Biology Subject Test terms—in the way that leads you to quickly choose correct answers on the Biology Subject Test. *That's* what raises your score and that's what we provide.

Maybe you're wondering: "What are these topics that are almost sure to appear on the Biology Subject Test? What subjects are you going to teach me?" Well, if you're curious we'll be glad to list them for you here and now. As you look at the list you might think it looks pretty much like a list of ordinary biology textbook topics. But remember this: The way *we* treat these topics is *not* the way your textbooks treat them. Our book is designed for one purpose only: to raise your Biology Subject Test score.

Among other things, we'll be talking about:

- Basic cell structure and organelles
- Aerobic cellular respiration: the Krebs cycle, the electron transport chain, and oxidative phosphorylation
- Anaerobic cellular respiration including fermentation
- Transmembrane transport: diffusion, active transport, facilitated transport
- Chromosomes, DNA, and RNA
- Nucleotides and base-pairing
- Transcription and protein synthesis
- Reproduction in humans, animals, and plants
- Mitosis, meiosis, fertilization, and embryonic development
- Bacteria, viruses, and fungi
- Basic chemistry and organic chemistry, including proteins and carbohydrates
- The human nervous system
- The human respiratory system
- The human heart and circulatory system
- Human blood
- The human endocrine system
- The human digestive system
- Genetics and Mendel's laws

- Evolution and phylogeny
- Animal learning and behavior
- Ecology and biomes
- Nutrition in plants and animals

And lots of other things that come up along the way.

STRATEGY 2: PRACTICE THE RIGHT *THING* AT THE RIGHT *TIME*

The list we just made for you does *not* represent the *order* in which we teach things. We've devised an order that's tailored to our purpose: raising your Biology Subject Test score.

Now look at the list. Do we address absolutely *all* of biology? No (although we cover quite a lot). It would be a waste for us to teach you the whole of basic biology. Our job isn't to teach you biology, but to raise your score on the Biology Subject Test—and those two things are *not* the same. Learn what we teach you and you'll certainly raise your score.

Chapters 3, 4, and 5 of this book present you with interactive questions all along the way—so you can learn *immediately* to take the knowledge and techniques we show you and apply them directly to the Biology Subject Test. Chapter 6 provides a full-length simulated Biology Subject Test for final practice.

STRATEGY 3: EASY STUFF FIRST

In all three sections of the regular Biology Subject Test; A, B, and C, and through the core sections of the E/M test the easier questions usually come earlier and the harder ones, later. It makes sense to begin a section and answer as many questions as you can until they start to become a little more difficult. Then go to the next section and do the same. Once you've answered the relatively easy questions in all the sections, go back to each section and start answering the harder ones (although after reading this book, hopefully you won't find too many of them very hard).

All questions carry the same credit. The ETS scoring machines don't know an easy question from a hard one. Answering a hard question correctly doesn't do you one more bit of good than answering an easy one. So don't waste time on harder questions when you could be answering easier ones. Don't attack the test in the order of its numbered questions, but in the order of its difficulty.

STRATEGY 4: TAKE A GUESS!

If you can actually eliminate some wrong answer choices, guessing will definitely help you. That's why Chapters 3, 4, and 5 of this book teach strategies that help you eliminate wrong answer choices. Use them!

After all, if you know that 2 out of 5 answer choices are wrong, you've got a one-third chance of guessing correctly from those that remain. That makes a big difference in your score. If you know that 3 out of 5 answer choices are wrong, you have a 50/50 chance of guessing right. That makes a *very* big difference in your score.

Otherwise, if you have no idea what you're doing, leave the question blank. You can leave a number of questions blank and still do well. For instance, on the regular test a raw score of 60 to 63 will give you a score of 600.

With that in mind, we're next going to talk about a strategy called:

Strategy 5: I, II, III — You're Out!

As we said, some questions will look like this:

- Which of the following is (are) . . . ?
    - I. blah, blah, blah . . .
    - II. blah, blah, blah . . .
    - III. blah, blah, blah . . .

  (A) I only
  (B) II only
  (C) I and III only
  (D) II and III only
  (E) I, II, and III

When you see questions like these, think logically about throwing out wrong answers even if you don't know the right one.

Here's how you do that. Forget biology for a minute and look at this question:

- Among the following, which is (are) ordinarily served as a dessert?
    - I. Fish filet
    - II. Pastry
    - III. College ice

  (A) I only
  (B) II only
  (C) I and II only
  (D) II and III only
  (E) I, II, and III

Never heard of "college ice"? Doesn't matter. You can still use strategy to raise your score.

You know fish (option I) isn't dessert. Eliminate all answers that mention it. A, C, and E are gone, which means you're left with B and D. Think what that means. You've got a *50/50 chance* of guessing correctly. (Just for your information, "college ice" is an old term for "ice cream sundae," so the correct answer is **D**.) Do that with ten questions and you'll get about five of them right. The scoring machine at ETS doesn't know *why* you get them right. It just marks your answers correct, and adds the credit to your score. So, when it comes to the I, II, III question type, use this strategy. It's a real score-raiser.

Strategy 6: Avoid the Camouflage Trap

In A-type questions, the Biology Subject Test writers sometimes try to make you fall into something we call the "camouflage trap."

Read these two sentences:

1. When two populations of the same species are separated for hundreds of thousands or perhaps millions of years, the diversity of their environments produces genotypic and phenotypic changes whose ultimate result is to produce such evolution as creates reproductive separation and hence, two or more distinct species.

2. If one species is divided into two groups, each then being sequestered in a separate set of living conditions and deprived of contact with the other, each group will adapt to its environment, and the natural selection process will so alter its chromosomal characteristics and gene pool as to create a situation ultimately in which the two separated groups can no longer interbreed, meaning that one or more new species will emerge.

These two statements mean exactly the same thing, but their wording is very, very different. Many of the words and phrases in statement 2 have the same meaning as those in statement 1, but they're disguised—they're *camouflaged*:

When two populations of the same species are separated…	*is camouflage for*	If one species is divided into two groups . . .
…the diversity of their environments produces genotypic and phenotypic changes…	*is camouflage for*	…each group will adapt to its environment, and the natural selection process will so alter its chromosomal characteristics . . .
…produce such evolution as creates reproductive separation…	*is camouflage for*	…so alter its gene chromosomal characteristics and gene pool as to create a situation ultimately in which the two separated groups can no longer interbreed…

So What About It?

When you learn something, whether it's biology or anything else, you usually learn it with certain *words* in mind. For instance, maybe you think of an "enzyme" as a "protein that acts as an organic catalyst." Okay, that's fine. But suppose you're totally attached to *those* words. Think what will happen if you get an A-type question like this:

- Which of the following would best describe an enzyme?
  (A) An organic substance that increases the rate law of a biochemical reaction
  (B) A polypeptide tending to reduce activation energy for a biological reaction
  (C) A protein that shifts dynamic equilibrium to the right
  (D) An organic molecule that is consumed during a biochemical equilibrium in order to favor the forward reaction
  (E) A carbon-containing compound that controls the rate at which reversible reactions achieve equilibrium

If you're too attached to the way that *you* usually think about enzymes you might not see the right answer even though (a) you know it, and (b) it's staring you in the face. The right answer to this question is **B**. Protein means almost the same thing as polypeptide, and organic catalyst means, basically, something that reduces activation energy for a biological reaction. The words aren't the same, but the meaning *is* (pretty much, anyway).

Many students who do know what an enzyme is don't answer this question correctly because they're used to thinking about it in words that don't appear in choice B. They look quickly through the choices. They see nothing they recognize and go into answer-choice panic. They pick something that sounds right—something that has the phrase "organic molecule" in it, like choice D. That's too bad, because students who know their biology nonetheless choose incorrect answers. Why? Because they fall into the camouflage trap.

Here's another example. Suppose you know that an ovum is a "gamete." Suppose you're married to *that* way of thinking about it. Think what will happen when you see this A-type question:

- Which of the following correctly describes an ovum?
    I. It secretes luteinizing hormone.
    II. It is produced by a gonad.
    III. It is a monoploid cell.

  (A) I only
  (B) I and II only
  (C) I and III only
  (D) II and III only
  (E) I, II, and III

Let's say you know what an ovum is. (You will know what it is after you read Chapter 4 of this book.) You know (1) it's produced by the female's ovary, and (2) the male's sperm fertilizes it to form a zygote. You even know that an ovum is haploid, not diploid.

You know *all* of that, but you still have trouble with this question. Why? Because you're not *used* to thinking of an ovum as options I, II, and III describe it.

monoploid	*is camouflage for*	haploid
gonad	*is camouflage for*	ovary

You know your biology, but you might not know to answer this question with D. Why? Because you fell into the camouflage trap. The camouflage trap is bad news.

HERE'S THE GOOD NEWS: YOU CAN AVOID THE CAMOUFLAGE TRAP

To avoid the camouflage trap, keep some simple rules in mind:

1. Don't take a test with blinders on your brain.
2. Remember there's more than one way to say the same thing.
3. When you face a question and know very well that it's something you've studied and memorized, don't become unglued just because the right answer doesn't leap out at you.

4. Chill out. Realize that the right answer is probably camouflaged by words that are different from the ones you have in mind. Search for them *calmly* and all of a sudden they *will* leap out at you.

Don't expect test writers to use *your* words. You might express an idea in one way and they might express it in another. Keep in your mind the *concepts* you know. Don't get too attached to the *words* that you usually use to express them.

Strategy 7: Avoiding the Temptation Trap

Suppose we gave this question to a seven-year-old child.

- Which of the following best characterizes a simple carbohydrate?
  (A) It is metabolized without enzymes.
  (B) It serves as an organic catalyst.
  (C) Columbus discovered America.
  (D) It is the product of multiple peptide bonding.
  (E) It is a molecule containing only carbon, hydrogen, and oxygen in the ratio 1:2:1, respectively.

The child won't know what any of this means but she probably will know that Columbus discovered America. So, not knowing what else to do, she chooses it; it's something she happens to know, so it seems right. (The right answer happens to be **E**. You'll learn about carbohydrates in the next chapter.)

The test writer took something the student knows and stuck it in among the answer choices. To the seven-year-old child, choice **C** is a familiar face, so she goes for it even though it has nothing to do with the question. When she fills in the little oval marked **C**, she falls into the temptation trap.

What's That Got to Do with Me and the Biology Subject Test?

A lot. On the day you take the test, there will be lots of things you know and some things that you don't. When you meet up with a tough question and you think you're lost, you might look for a familiar face—you might grab an answer that makes a statement you've heard before.

Let's suppose you know all about the human heart and circulation. You know that

- most arteries carry oxygen-rich blood
- most veins carry oxygen-poor blood
- the pulmonary artery carries oxygen-poor blood from the right side of the heart to the lungs
- the pulmonary veins carry oxygen-rich blood from the lungs to the left side of the heart

(If you don't know these facts now, don't worry—you will.)

Now look at this question:

- A sample of blood is taken from an unknown site in a human patient. The blood shows an oxygen content equivalent roughly to that of the venous and not the arterial circulation. Among the following, which statement best applies to the blood that was drawn?

  (A) It was drawn from the pulmonary vein and is rich in oxygen.
  (B) It was drawn from an alveolus and is rich in oxygen.
  (C) It was drawn from large branches of the pulmonary artery.
  (D) It was drawn from the anterior vena cava and will enter the heart at the left ventricle.
  (E) It was drawn from the left ventricle and will leave the heart via the aorta.

If the right answer to this question doesn't leap out at you, you might decide to pick choice E because it rings true. Blood that leaves the left ventricle does enter the aorta. So E *sounds* right. But E is wrong because *it doesn't answer the question*. If you choose it, you're like the seven-year-old child who chooses the statement about Columbus. You fall into the temptation trap, and that's bad.

But Here, Again, Is the Good News: You Can Avoid the Temptation Trap Too

When you find yourself running toward an answer just because it makes a familiar statement—stop and think. Realize that you may be falling into the temptation trap. Look at the *question* carefully. Then, look at the answer choices to see if one of them really *does* answer the question. And while you're looking, keep an eye out for camouflage.

You see, the temptation and camouflage traps often go together. One choice tempts you while some other choice—the right one—is sitting there in camouflage.

Look again at the blood question. It talks about blood that has an oxygen content similar to that of the venous circulation. That means it's oxygen poor. You're tempted to think it comes from some vein. The trouble is: there's no answer choice that says that exactly.

But look at choice C. It's saying that the blood comes from large branches of the pulmonary artery. The pulmonary artery carries blood from the right ventricle to the lungs. As you'll learn, it's the only artery in the human body that carries oxygen-poor blood. The large branches of the pulmonary artery also carry blood that's oxygen poor. Even though they're branches of an artery, their oxygen content resembles that of the venous circulation.

The right answer is **C**, and you knew that, but you might have chosen something else. Why? Because of the temptation trap. So, when a question makes you feel lost in a storm, don't rush to the first shelter that looks safe. There's probably a bomb inside. Look, instead, just behind the bush, or under the rock. The safety you need is probably hiding there—in camouflage.

STRATEGY 8: MAKE FRIENDS WITH THE ENEMY

When it comes to tests, too many people think too much about finding answers. They forget that the answer has to fit the question.

Here's what we mean. Put biology aside again, just for a minute, and look at this math question:

- If the third digit in the base ten number system is multiplied by a number equal to a number that is no more than and no less than the sum of the first and second digits in the base ten number system, and the numerical result of that calculation is then divided by that single digit in the base ten number system of highest numerical value, the final result will be

    (A) 1.0
    (B) 1.5
    (C) 2.0
    (D) 2.5
    (E) 3.0

If you're like most students, you're so horrified by the way this question is worded, you'd rather do anything than try to understand it. You'll roam around the answer choices A through E, and as though you were shopping for underwear you'll just decide on "that one," whether it's A, B, C, D, or E.

The point is that the question is so difficult to read, you'd rather forget about it. You pretend you're thinking about the answers, deciding which one somehow looks best, and without even *knowing what the question is, you try to answer it*. That's a natural reaction, but it's no way to take a test.

You absolutely *do* know the answer to this question because the question is simple. You won't know it's simple, however, unless you take it apart and translate it into plain English. So, let's do that. Let's look at the question piece by piece and find out what it *is*.

"If the third digit in the base ten number system . . ."

◆ that's 3, so we're now talking about the number 3.

". . . is multiplied by . . ."

◆ we're going to multiply 3 by something. By *what*? Well, let's keep reading.

". . . a number equal to . . ."

◆ we're going to multiply it by a number that's equal to something. Equal to *what*? Let's keep reading.

". . . a number that is no more than and no less than . . ."

◆ we're going to multiply 3 by a number that is no more than and no less than something. "No more than and no less than" means "equal to," doesn't it? So still, we know that we're going to multiply 3 by some number. What number? Keep reading.

> "... the sum of the first and second digits in the base ten number system ..."

- the number is the sum of 1 and 2, which is 3. In other words we now know we're going to multiply 3 by 3 (3 × 3), which is equal to 9. We're beginning to get somewhere.

> "... and the numerical result of that calculation ..."

- Ah? We're going to do something with that 9. What are we going to do with it? Keep reading.

> "... is then divided by ..."

- We're going to divide it by something. By what? Keep reading.

> "... that single digit in the base ten number system ..."

- We're going to divide it by some digit in the base ten number system. What digit? Keep reading.

> "... of highest numerical value ..."

- We're going to divide 9 by the highest digit in the base ten number system, which happens to be 9. So, we're going to divide 9 by 9—(9 ÷ 9).

> "... the final result will be ..."

- Guess what? Now we know what the question says. It tells us to: take 3, multiply it by 3, and divide the result by 9.

Now, if we go looking at the answer choices, we know what we're looking for. We're looking for 1. And what do you know? It's there—choice **A**.

WHAT EXACTLY WAS THE POINT OF THAT?

When you take tests, there are probably lots of times that you start deciding on an answer before you know what you're being asked. Why do you do that? Because the question is long or confusing or badly written, and it's just too much trouble to bother with. So, you actually try to answer a question without knowing what the question is. That's no way to do things.

When you find a question to be hard, ask yourself:

1. Am I failing to understand what the question is?

2. Is that why I think it's hard?

If the answers are yes, then you're *not* really finding the question hard, and you probably do know the answer. You just don't know what the question is. So what do you do? You swallow hard, grit your teeth, and get to know the question. Tear it apart piece by piece. Figure out what it's asking. Then go to the answer choices, and the right answer will probably pop out at you. As you move through Chapters 3, 4, and 5, you'll see just how this works.

STRATEGY 9: KEEP BOTH EYES OPEN

Some of the Biology Subject Test questions require that you take two items of knowledge and put them together. Let's leave biology and talk about bicycles.

- You're riding a bicycle and are about to crash into something. You know:
  (1) the best thing to do is stop
  (2) the brakes will help you do that
  (3) you use the brake by squeezing the handles that sit just under the handlebars on each side

So there are *three* items of knowledge that help you out of this jam. Suppose, then, that we ask you this question:

- When a bicyclist finds she is about to collide with a pedestrian, her best course of action is to
  (A) proceed with no change
  (B) increase velocity rapidly
  (C) close her eyes
  (D) increase the force she exerts on the pedals
  (E) squeeze the two short movable handles located just under each of the handlebars

The answer, obviously, is **E**. But notice that in order to choose E, you must put together (without realizing it maybe) three separate bits of knowledge:

1. that the bicyclist should stop
2. that the brakes will help her do it
3. that the brakes are activated by those short little handles under the handlebars

This same sort of thing comes up on the Biology Subject Test. Look again at this question. (But don't try to answer it yet.)

- An experimenter subjects an aerobic bacterium to oxygen-rich conditions between times 1 and 2 and then to oxygen-poor conditions between times 2 and 3. Which of the following changes will most likely take place between the two time periods?
  (A) Oxygen consumption will increase.
  (B) ATP production will cease.
  (C) Carbon dioxide production will increase and ATP production will increase.
  (D) Carbon dioxide production will increase and lactic acid production will decrease.
  (E) ATP production will decrease and lactic acid production will increase.

In order to answer this question, you have to know *four* things:

1. that both anaerobic and aerobic respiration produce ATP
2. that aerobic respiration produces more ATP than does anaerobic respiration
3. that an aerobic organism can perform anaerobic respiration if it has to
4. that anaerobic respiration produces lactic acid

Once you realize that the question is asking you to think about *four things at once*, you can easily find the answer. If the question were broken into the following four questions, you probably wouldn't have trouble with them.

1. What do anaerobic and aerobic respiration have in common?
    (A) Both produce ATP.
    (B) Both produce acetic acid.
    (C) Both require oxygen.
    (D) Both involve the use of Krebs cycle enzymes.
    (E) Both require the presence of hemoglobin.

—The answer is **A**.

2. In comparison to anaerobic respiration, aerobic respiration
    (A) does not require mitochondria
    (B) does not require oxygen
    (C) produces greater amounts of lactic acid per molecule of glucose
    (D) produces smaller amounts of ATP per molecule of glucose
    (E) produces greater amounts of ATP per molecule of glucose

—The answer is **E**.

3. Which of the following is true concerning aerobic organisms?
    (A) They do not use ATP as their energy currency.
    (B) They cannot function in the presence of oxygen.
    (C) Their cells lack mitochondria.
    (D) In the absence of oxygen they will undergo anaerobic respiration.
    (E) They tend to be small desert animals.

—The answer is **D**.

4. Among the following, which is produced during anaerobic respiration?
    (A) ADP
    (B) Lactic acid
    (C) Carbon monoxide
    (D) Protein
    (E) Sucrose

—The answer is **B**.

You may have known all four of these answers. (If you didn't, you will by the time we're through with you.) Well, if you can answer all of these questions then you can answer a question that mixes them into one. But in order to *know* that you can answer it, you've got to realize that the question writer is really asking four questions—not just one. So, when you're up against a question that seems to be tough think that maybe the question seems tough just because there's more to it than meets the eye. Two or more questions might be blended into one.

LET'S GET GOING ALREADY!

Now, we'll get on with Chapters 3 through 5, in which we:

1. review the biology that's sure to show up on the Biology Subject Test

and

2. test you along the way with interactive questions that make sure you're learning and reviewing biology in the way the Biology Subject Test tests it—*in the way that's going to raise your score.*

So, turn the page and let's begin Chapter 3: Cracking Cellular and Molecular Biology.

3

Cracking Cellular and Molecular Biology

KNOW ANYTHING ABOUT ORGANIC CHEMISTRY?

The Biology Subject Test writers say they test a little organic chemistry, but the stuff they test is pretty simple. So if you haven't taken organic chemistry, don't worry. We'll teach you all the organic chemistry you'll have to know.

Atoms are the fundamental stuff of the physical world. They get together in chemical reactions and form molecules.

$$H + H \to H_2$$

Molecules can also react with atoms or other molecules to form molecules.

$$2H_2 + O_2 \rightarrow 2H_2O$$

A compound is just a molecule with *different* types of atoms in it. For example, CCl_4 is a compound because the molecule has both carbon and chlorine in it. On the other hand, H_2 is a molecule, but it's **NOT** a compound. That's because the only atoms within the molecule are hydrogens.

When we think about chemical reactions we talk about the reactants (on the left side) and the products (on the right side.) Got it all? Okay. Fill in the appropriate boxes.

- H_2O [❏ is ❏ is not] a *compound*.

- Cl_2 [❏ is ❏ is not] a *compound*.

- Cl_2 [❏ is ❏ is not] a *molecule*.

What *Organic* Chemistry Means

When the Biology Subject Test talks about "organic chemistry," it's talking about molecules that have *carbon* in them. Organic molecules have carbon in them. Inorganic molecules don't.

Fill in the appropriate boxes.

- Water (H_2O) is an [❏ **organic** ❏ **inorganic**] compound.

- Methane (CH_3) is an [❏ **organic** ❏ **inorganic**] compound.

For the Biology Subject Test you have to know something about:

1. amino acids and proteins
2. carbohydrates and
3. lipids

So, here goes.

ORGANIC CHEMISTRY LESSON 1: AMINO ACIDS AND PROTEINS

First of all, an amino acid is a molecule. There are lots of different amino acids, but here's one we might call typical.

$$R - \underset{\underset{NH_2}{|}}{\overset{\overset{H}{|}}{C}} - \overset{\overset{O}{\|}}{C} - OH$$

We've put a box around the backbone of the amino acid. It's the part of the molecule to the right of the letter R. Look at the structure in the box. It's got two carbons. One carbon has an H and an NH_2 bonded to it. The NH_2 group is called the *amino* group. The other has an O and an OH bonded to it. The O is bonded to the second carbon with a *double* bond. The COOH group is called the carboxyl group. If you know what the boxed structure looks like, with its amino and carboxyl group, you'll be able to recognize amino acids on the Biology Subject Test. So take a long look at that structure in the box, and then draw it below, *three times*, so you really know what it looks like.

- Draw it:

- Draw it again:

- Draw it again:

Good work. Now you know about the backbone of the amino acid—the part you've learned to draw.

But what about the R part of the molecule? The R part of molecule makes each amino acid different from all others. *All* amino acids have the structure that's in the box, but different amino acids differ with respect to R. R could be anything from a simple hydrogen atom to a whole long chain of carbon atoms with all sorts of things bonded to them. The R part of the molecule gives the amino acid its identity.

In an amino acid called glycine, R is just a hydrogen atom:

$$\begin{array}{c} H \quad O \\ | \quad \| \\ H-C-C-OH \\ | \\ NH_2 \end{array}$$

In an amino acid called cysteine, R has some carbon and some sulfur too.

$$\begin{array}{c} \text{Sulfur} \\ \downarrow \\ H-S-C-C-C \\ \end{array}$$
Cysteine

AMINO ACIDS GET TOGETHER TO FORM PROTEINS

When a whole bunch of amino acids get together, they form an amino acid chain. An amino acid chain of sufficient length is *a protein*. Let's see how *two* amino acids get together. Here are two amino acids.

We've drawn circles around the **OH** on one amino acid and an **H** on the other.

The carbon from one amino acid loses the OH and it bonds instead to the nitrogen on the other amino acid. Meanwhile the nitrogen on the other amino acid loses an H.

peptide bond

The bond between two amino acids is called a *peptide bond*. Notice what gets removed—H_2O. That's why peptide bonds are said to be formed by dehydration synthesis.

When a whole bunch of amino acids link together and form a long amino acid chain, you get a protein. A protein is sometimes called a **polypeptide.**

Reread what we just told you about the formation of proteins, and fill in the blanks.

- A protein is a chain of _____ _____.

- The bond that links two amino acids together is called a _____ bond.

- Because proteins are, essentially, chains of amino acids linked together by _____ bonds, a protein might also be called a _____.

We've discussed the way in which peptide bonds are formed. The test writers might also ask about the way in which they're *broken*. That's no problem. The picture we've drawn above shows you that a peptide bond is formed when a molecule of water (H_2O) is removed during dehydration synthesis. That process is sometimes called condensation. The *breakage* of a peptide bond involves the *addition* of a molecule of water, and that process is called hydrolysis:

- The process in which two amino acids are separated involves the breakage of a _____ bond.

- The breakage of a peptide bond requires the addition of _____ and is given the name _____.

- The disassembly of a polypeptide into its component amino acids involves the [❏ **addition** ❏ **removal**] of water, and is called _____.

- The assembly of a polypeptide from its amino acid constituents involves the [❏ **addition** ❏ **removal**] of water, and is called _____.

ORGANIC CHEMISTRY LESSON 2: CARBOHYDRATES

When we talk about carbohydrates, we mean a whole class of molecules. However, all carbohydrates have *this* in common: They're made only of carbon, oxygen, and hydrogen. For any carbohydrate, the number of carbons is equal to the number of oxygens, and the number of hydrogens is equal to twice the number of either carbons or oxygens. In other words, the Cs, Hs, and Os in most carbohydrates exist in a 1:2:1 ratio:

$$C_nH_{2n}O_n$$

When we discuss carbohydrates we end up using terms like: monosaccharides, disaccharides, polysaccharides, and hexose. Fortunately, however, these words are easy to understand. When we say "monosaccharide," we're talking about carbohydrates that are pretty small (you can remember this because *mono-* means *one*, and one is a pretty small number.)

Actually, the only monosaccharides the Biology Subject Test really cares about are:

1. glucose and
2. fructose.

Glucose and fructose are both carbohydrates. More specifically, they're monosaccharides. Glucose and fructose *have the same chemical formula*: $C_6H_{12}O_6$.

So What's the *Difference* Between Glucose and Fructose?

Glucose and fructose differ in the *arrangement of their molecules*. Take a look:

Glucose

Fructose

How do the two molecules differ? In the glucose molecule, the double-bonded oxygen is located on the top carbon. In the fructose molecule, it's located on the second carbon down.

For the Biology Subject Test, that's really all you have to know about the difference between glucose and fructose:

- Both glucose and fructose are carbohydrates, and both are monosaccharides.
- Both have the formula $C_6H_{12}O_6$.
- Glucose and fructose *differ* in the way the double-bonded oxygen is oriented within the molecule.

Look at the picture of glucose and fructose on page 28. Remember it and then cover it up. Look at the picture below and then fill in the blanks and appropriate boxes.

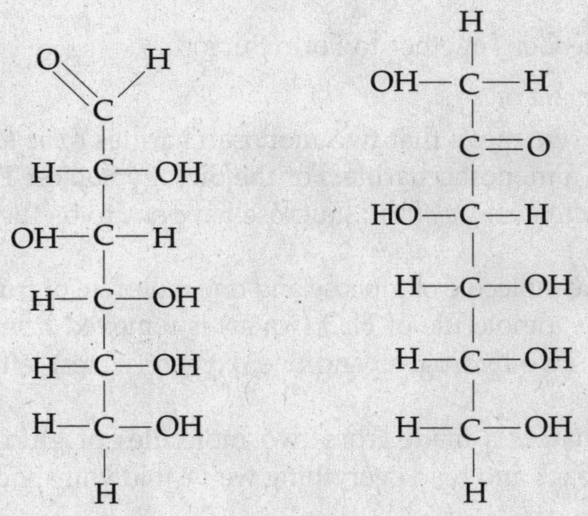

Molecule A

Molecule B

- Molecule A is called _____.
- Molecule B is called _____.
- The chemical formula for glucose is _____.
- The chemical formula for fructose is _____.
- Glucose and fructose [❏ **are** ❏ **are not**] identical molecules.
- Glucose and fructose [❏ **do** ❏ **do not**] have identical chemical formulas.

Glucose can also form a ring.

Glucose and Fructose Get Together to Form Sucrose, Which is a *Disaccharide*

When we say disaccharide, we mean that two monosaccharides *come together* and form a molecule that's about twice as big as a monosaccharide. For the Biology Subject Test, some disaccharides you want to know about are *sucrose* and *maltose*. (Sucrose happens to be the stuff we call "sugar" when we're talking plain English.)

Sucrose is made from one molecule of glucose and one molecule of fructose. When the glucose and fructose molecules combine, a molecule of H_2O (water) is removed from the scene. The formula for sucrose **IS NOT** $C_{12}H_{24}O_{12}$. Two hydrogens and one oxygen (water) left. The formula for sucrose is $C_{12}H_{22}O_{11}$.

Maltose, on the other hand, is made from two molecules of glucose. It is also produced via dehydration synthesis. Go back and read everything we've told you about carbohydrates. Then fill in the blanks.

- A molecule of maltose is formed from two molecules of _____ .

- A molecule of glucose and a molecule of fructose, both of which are _____saccharides, combine to form a molecule of _____, which is a _____ saccharide.

- The chemical formula for both glucose and fructose is _____.

- The chemical formula for sucrose is _____.

KNOW ABOUT SOME POLYSACCHARIDES

When lots and lots of glucose molecules link together in a large chain, they can form (1) glycogen (2) cellulose or (3) starch. All of these are long chains of glucose molecules, and all are called polysaccharides. (There are other polysaccharides in this world, made from other monosaccharides, but you don't have to worry about them.)

Now, if glycogen, cellulose, and starch all come from large chains of glucose molecules, what's the difference between them? The difference between them is in the *way* the glucose molecules are linked together, and that's all you have to know about it.

ONE MORE THING TO KNOW ABOUT GLYCOGEN AND CELLULOSE

Since glycogen and starch are large chains of glucose, they act as a good storage form of glucose. In a way, each molecule is like a package of glucose. Think of glycogen and starch as molecules that *store* glucose. And—very important—remember this:

- Plants use **starch** to store glucose.
- Animals use **glycogen** to store glucose.

Review everything we just taught you about glycogen and cellulose. Then fill in the blanks and appropriate boxes.

- Large-chained molecules of glucose create molecules of _____, _____, or _____.

- Cellulose is a _____ saccharide.

- Glycogen is a _____ saccharide.

LIPIDS

Lipids are also made up of carbons, hydrogens and oxygen, but not in a 1:2:1 ratio. Lipids contain three *fatty acids* and one *glycerol* molecule. What's a fatty acid? It's a hydrocarbon chain (a string of carbons with hydrogens attached to them) that has a carboxyl group at one end. What's a glycerol molecule? Well, for one thing it's an alcohol that has three carbons in it. Lipids also undergo dehydration synthesis and hydrolysis. Some examples of lipids are fats and oils.

- Cellulose and glycogen differ in the way that _____ molecules are bonded together.

- Starch serves as a means for storing glucose in [❑ **plants** ❑ **animals**].

- Glycogen serves as a means for storing glucose in [❑ **plants** ❑ **animals**].

- Lipids are composed of three _____ and one _____ molecule.

So much for any organic chemistry you might see on the Biology Subject Test. Let's go to another subject.

CHEMISTRY AND THE ATMOSPHERE: THE HETEROTROPH HYPOTHESIS

Since life depends on the Earth's atmosphere, the test writers might want you to know something about its chemical contents. Earth's atmosphere *today* contains 78% nitrogen, 21% oxygen and a tiny amount of carbon dioxide. (It also has teensy amounts of helium and neon.)

Life first showed its face on this planet around one billion years ago. The atmosphere was different from what it is today. According to the heterotroph hypothesis life began in an atmosphere that did *not* contain much oxygen. Instead, it contained lots of hydrogen (H_2), ammonia (NH_3), methane (CH_4), and water (H_2O).

So remember:

Today's atmosphere contains mostly nitrogen and oxygen

Earth's early atmosphere, in which life began, contained mostly hydrogen, ammonia, methane, and water; it did NOT contain oxygen.

- At *present*, the gas of highest concentration in the Earth's atmosphere is [❏ helium ❏ oxygen ❏ nitrogen].

- According to the heterotroph hypothesis, oxygen [❏ was ❏ was not] a chief component of the atmosphere when life first began.

- According to the heterotroph hypothesis, hydrogen [❏ was ❏ was not] a chief component of the atmosphere when life first began.

- According to the heterotroph hypothesis, methane [❏ was ❏ was not] a chief component of the atmosphere when life first began.

- According to the heterotroph hypothesis, ammonia [❏ was ❏ was not] a chief component of the atmosphere when life first began.

- According to the heterotroph hypothesis, water [❏ was ❏ was not] a chief component of the atmosphere when life first began.

HERE'S SOMETHING YOU'VE SEEN BEFORE: A EUKARYOTIC CELL

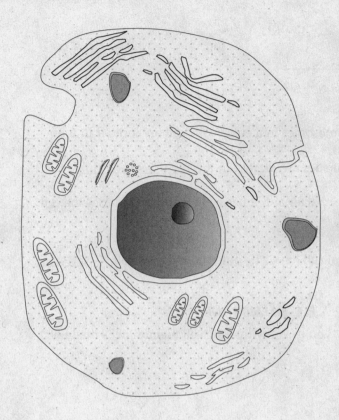

As you probably remember, every living thing—plant or animal—is made of cells. And by the way, for the Biology Subject Test we don't say "living thing." We say "organism." So, all *organisms* are made of cells, and according to the cell theory, (which is certainly correct) all cells arise from preexisting cells. The simplest organisms, like amoebas, are made of one cell. Complex organisms like maple trees, mango plants, snakes, and human beings are made up of lots and lots of cells. They're multicellular.

Look back at the picture, and think of a cell as having three areas:

1. *the cell wall and membrane*, which form the outer portion,

2. *the cytoplasm*, in which the organelles exist, and

3. *the nucleus*, which is bounded by a nuclear membrane and in which the chromosomes are located.

We just said that cells have as their outermost layers a cell wall and a cell membrane. It turns out, however, that only plant cells (and fungus cells) have cell walls. Animal cells don't. Here, then, is the whole truth: A plant or bacterial cell has as its outermost layers a cell wall and a cell membrane. An animal cell has a cell membrane, but *not* a cell wall.

Go back a few lines. Read again about the three areas of a cell and what's inside them. Then fill in these blanks.

The cell's three areas are the:

- 1. _____ in plants and bacteria and _____ in animal cells.

- 2. _____, and

- 3. _____.

In particular,

- The chromosomes in a eukaryotic cell are located in the _____.

- The organelles are located in the _____.

- A plant or bacterial cell has as its outer boundaries the cell _____ and the cell _____ .

- An animal cell has as its outer boundary a cell _____.

MORE ABOUT THE CELL MEMBRANE

The test writers want you to know that the cell membrane is made of:

1. protein

   and

2. lipid.

We learned about lipids when we covered organic chemistry a few pages back. Remember that the cell membrane is made of both protein and lipid.

The test writers also want you to know something about how things get through the membrane and into the cell. There are four ways that molecules make their way through the cell membrane and into the cell. Don't try to understand what we're going to tell you about them. Just remember it on test day.

FOUR WAYS THAT STUFF GETS INTO CELLS

Way 1: Passive diffusion. When there's a relatively high concentration of something outside the cell and a relatively low concentration inside the cell, the stuff just passes right in as though it were moving through a door. Basically, the stuff (whatever it is) relieves the crowding outside the cell by moving some of itself inside the cell.

But keep this in mind

The cell membrane is made, partly, of lipid. Passive diffusion only works if the stuff that wants to get into the cell is soluble in lipid (lipid soluble). Passive diffusion only works for lipid-soluble substances.

Way 2: Facilitated diffusion. Facilitated diffusion lets various molecules get past the cell membrane and into the cell even if they're *not* lipid soluble. Facilitated diffusion allows the cell to take in substances that are *not* lipid soluble.

How does facilitated diffusion work? The substance that wants to get into the cell attaches to some substance that *is* lipid soluble. That lipid-soluble substance is the carrier molecule. The carrier molecule takes the lipid *in*soluble substance right across the membrane and into the cell. Once the carrier and its lipid passenger are inside, they separate.

And keep this in mind, too

Facilitated diffusion is like passive diffusion in one important respect: A molecule gets into the cell only if its concentration outside is higher than its concentration inside. Facilitated diffusion and passive diffusion also apply to substances moving from the inside of the cell to the outside. In this direction, the concentration of the substance must be higher inside the cell than it is outside the cell. When the Biology Subject Test mentions facilitated diffusion or facilitated transport think:

Carrier molecule takes lipid insoluble substance across the cell membrane. Movement only occurs from an area of higher concentration to an area of lower concentration.

Way 3: Active transport. When we say active transport, we're saying that the cell has to *expend energy* to take something from outside the cell, move it across the membrane, and get it inside the cell. Active transport is different from passive diffusion and facilitated diffusion in this important respect: It can move a substance across the cell membrane *even if the substance's concentration is greater inside than outside.* Here's another way to say that: Active transport allows movement against a concentration gradient.

In fact, it's *because* active transport involves the expenditure of energy that it allows the cell to move material against a concentration gradient—to take in stuff that's already got a high concentration inside the cell and a low concentration outside the cell.

When it comes to active transport, no one will ask you about whether the stuff that's moving across the membrane is or is not lipid soluble, so don't worry about that. When the Biology Subject Test says active transport, think:

The cell has to expend energy, and it can move substances into or out of the cell against a concentration gradient.

Way 4: Endocytosis. When we say endocytosis, we think of the cell taking in some particle by engulfing it within a pocket we call a vacuole.

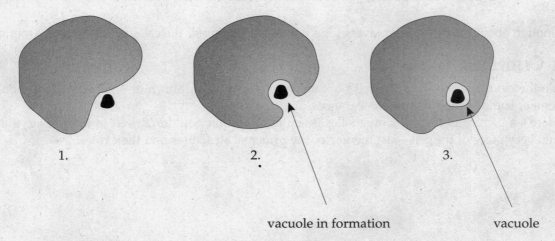

1.   2.   3.

vacuole in formation   vacuole

When you read endocytosis, think of a particle engulfed in a vacuole.

Go back to the point at which we started talking about cell walls and cell membranes. Reread it. Learn it. Then fill in the blanks and boxes appropriately.

- As its outermost layers, the [❏ **plant** ❏ **animal**] cell has *both* a cell wall and a cell membrane.

- As its outermost layer, the [❏ **plant** ❏ **animal**] cell has *only* a cell membrane and no cell wall.

- A cell membrane is made of _____ and _____.

- The process of passive diffusion [❏ **does involve** ❏ **does not involve**] the expenditure of energy.

- In the process of facilitated transport, a substance crosses the cell membrane with the help of a _____ molecule.

- In the process of facilitated transport, the substance that crosses the cell membrane [❏ **must be** ❏ **need not be**] soluble in lipid.

- The process of _____ _____ requires the expenditure of energy.

- In the process of active transport, movement [❏ **does** ❏ **does not**] proceed against a concentration gradient.

- In the processes of passive diffusion and facilitated transport, movement [❏ **does** ❏ **does not**] proceed against a concentration gradient.

Enough about the cell's outer layers for a while. Let's go *inside* the cell and take a look around.

THE CYTOPLASM

Inside the cytoplasm we find the cell's organelles. The cell has a bunch of organelles. You should remember some of their names, appearances, and functions.

Here's a picture of the cell's organelles. Keep in mind that you don't need to know everything about every piece of the cell—just memorize the principal structures and their roles.

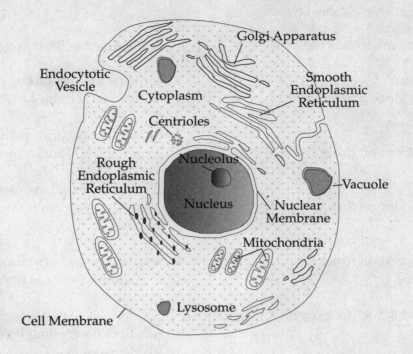

1. Vacuole—helps the cell expel waste.
2. Nucleus—contains genetic material which directs the activities of the cell.
3. Centrioles—help form spindle apparatus during mitosis.
4. Nucleolus—serves as site at which ribosomal RNA is formed.
5. Lysosome—digests foreign substances and worn organelles.
6. Chromosomes—contain genetic information (made up of DNA and protein).
7. Mitochondria—function in cellular respiration; formation of ATP.
8. Smooth endoplasmic reticulum—system of tubes that transport substances around the cell.
9. Rough endoplasmic reticulum—endoplasmic reticulum with ribosomes sitting on it. The ribosomes are the bumps, and they function in protein synthesis.
10. Golgi apparatus—packages and stores proteins for secretion out of cell, or targets them to specific membrane within cell.
11. Cell membrane—the outer membrane that regulates what goes in and out of the cell.
12. Ribosomes—sites of protein synthesis.

Study the picture and the list for a few minutes, and—we hate to say it—

MEMORIZE!

Then look at the organelles one by one, and for each, tell yourself (a) its name and (b) its function. After you've done that, take the quiz on the next page.

Complete the matching column below.

Organelle/Structure

Golgi apparatus:
Function/Description (from right) #_____

Chromosomes:
Function/Description (from right) #_____

Lysosomes:
Function/Description (from right) #_____

Centrioles:
Function/Description (from right) #_____

Nucleolus:
Function/Description (from right) #_____

Vacuoles:
Function/Description (from right) #_____

Mitochondrion:
Function/Description (from right) #_____

Nucleus:
Function/Description (from right) #_____

Cell membrane:
Function/Description (from right) #_____

Rough endoplasmic reticulum:
Function/Description (from right) #_____

Smooth endoplasmic reticulum:
Function/Description (from right) #_____

Ribosome:
Function/Description (from right) #_____

Function/Description

1. contains the chromosomes that control the activities of the cell

2. contain genetic information; composed of DNA and protein

3. digest foreign substances and worn organelles; contain hydrolytic enzymes

4. related, generally, to function in formation of spindle fiber during mitosis

5. function in expelling wastes; fluid-filled spaces

6. site at which ribosomal RNA is formed; located in nucleus

7. cellular respiration produces ATP; double-membraned with an inner matrix

8. cellular secretion; storage and removal of secretory products

9. consists of two lipid layers with proteins interspersed

10. sites of protein synthesis

11. channels with ribosomes

12. channels without ribosomes

Check out the answers on the next page to see how you did, and now take the test again.

Organelle/Structure	Function/Description

Golgi apparatus:
Function/Description (from right) #_____

Chromosomes:
Function/Description (from right) #_____

Lysosomes:
Function/Description (from right) #_____

Centrioles:
Function/Description (from right) #_____

Nucleolus:
Function/Description (from right) #_____

Vacuoles:
Function/Description (from right) #_____

Mitochondrion:
Function/Description (from right) #_____

Nucleus:
Function/Description (from right) #_____

Cell membrane:
Function/Description (from right) #_____

Rough endoplasmic reticulum:
Function/Description (from right) #_____

Smooth endoplasmic reticulum:
Function/Description (from right) #_____

Ribosome:
Function/Description (from right) #_____

1. controls the activities of the cell

2. contain genetic information; composed of DNA and protein

3. digest foreign substances and worn organelles; contain hydrolytic enzymes

4. related, generally, to function in formation of spindle fiber during mitosis

5. function in expelling wastes; fluid-filled spaces

6. site at which ribosomal RNA is formed; located in nucleus

7. cellular respiration produces ATP; double-membraned with an inner matrix

8. cellular secretion; storage and removal of secretory products

9. consists of two lipid layers with proteins interspersed

10. sites of protein synthesis

11. channels with ribosomes

ANSWERS

Organelle/Structure	Function/Description
Golgi apparatus: Function/Description (from right) # __8__	1. controls the activities of the cell
Chromosomes: Function/Description (from right) # __2__	2. contain genetic information; composed of DNA and protein
Lysosomes: Function/Description (from right) # __3__	3. digest foreign substances and worn organelles; contain hydrolytic enzymes
Centrioles: Function/Description (from right) # __4__	4. related, generally, to function in formation of spindle fiber during mitosis
Nucleolus: Function/Description (from right) # __6__	5. function in expelling wastes; fluid-filled spaces
Vacuoles: Function/Description (from right) # __5__	6. site at which ribosomal RNA is formed; located in nucleus
Mitochondrion: Function/Description (from right) # __7__	7. cellular respiration produces ATP; double-membraned with an inner matrix
Nucleus: Function/Description (from right) # __1__	8. cellular secretion; storage and removal of secretory products
Cell membrane: Function/Description (from right) # __9__	9. consists of two lipid layers with proteins interspersed
Rough endoplasmic reticulum: Function/Description (from right) # __11__	10. sites of protein synthesis
Smooth endoplasmic reticulum: Function/Description (from right) # __12__	11. channels with ribosomes
Ribosome: Function/Description (from right) # __10__	12. channels without ribosomes

More about the Cytoplasm: Ribosomes and the Endoplasmic Reticulum

If you look back at our picture of the cell, you'll see that one of the organelles is called smooth endoplasmic reticulum and another is called rough endoplasmic reticulum.

Generally speaking, the endoplasmic reticulum is a long, winding network of tubes that goes all over the cytoplasm carrying things from here to there. The endoplasmic reticulum that has bumps on it is rough endoplasmic reticulum. The endoplasmic reticulum that doesn't have bumps on it is smooth.

The bumps that appear on the rough endoplasmic reticulum are *ribosomes*. Ribosomes sit right on the outside of rough endoplasmic reticulum and make it look rough. What do ribosomes do? *Ribosomes are the sites at which the cell makes protein*. When you think of endoplasmic reticulum, think these things:

- Endoplasmic reticulum is a network of channels that runs throughout the cytoplasm.
- Rough endoplasmic reticulum is rough because it's got ribosomes sitting on it.
- Ribosomes are the sites of protein synthesis.

Got it? Fill in the blanks and boxes appropriately.

- The system of channels and tubes that run throughout the cytoplasm is called, generally, the _____ _____.

- The _____ _____ may be designated as _____ or _____.

- Within the cytoplasm, ribosomes are the sites of _____ _____.

- Within the cell, ribosomes are located on [❏ **smooth**   ❏ **rough**] _____ _____.

Now that we've talked about the organelles that are sitting around in the cytoplasm, let's talk about what goes on in the cytoplasm.

What Goes On in the Cytoplasm: Chemical Reactions and Enzymes

Mainly, cytoplasm is busy carrying on chemical reactions of all different types.

Here's a chemical reaction.

In this reaction, molecules X and Y are the reactants and Z is the product. Now get this: The reaction won't happen unless X and Y get together.

X and Y need a mutual friend who'll get them together in the same place at the same time so they can react and form Z. The mutual friend is an *enzyme*. Here's a picture of an enzyme that would work very nicely to get X and Y together. We'll call the enzyme molecule E. E's job is to catalyze the reaction between X and Y.

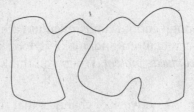

Molecule E

Notice that the spaces on the top of molecule E match up with molecule X's shape. The spaces on the bottom match up with molecule Y's shape.

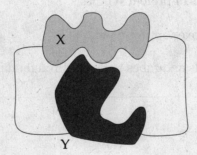

If molecule E is floating around in the cell, X and Y have a good chance of getting together. Once they do, they react to form molecule Z.

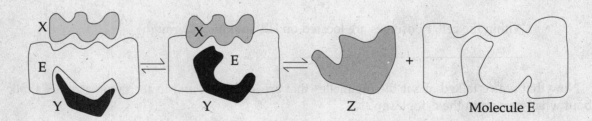

When the reaction is over, *molecule E is still there*. But its job here is done. In fact, it's free to go find *another* pair of X and Y and get them together to form another molecule of Z. So remember: When an enzyme (that's right, E stands for enzyme) catalyzes a reaction, the enzyme isn't used up. It's still there when the reaction is over. It can go find some more reactants and catalyze more reactions.

ENZYMES AND SOME IMPORTANT BIOLOGY SUBJECT TEST PHRASES

Sometimes when people talk about enzymes they say "lock-and-key theory." They mean that an enzyme and the products it's friendly with fit like a lock and key.

Some people say an enzyme is an organic catalyst. They just mean that enzymes help speed up *biological* reactions by getting the products together.

When we talk about enzymes we don't say reactant. We say substrate, instead. In the first reaction we looked at, molecule E was an enzyme and molecules X and Y were *substrates*. Molecule E catalyzed the reaction in which substrates X and Y formed product Z.

Here's another thing to keep in mind. Look at molecule E and the spots that are cut out for molecules X and Y. The cut out spots are called active sites. When we refer to an enzyme's active site, we mean the spot that's cut out for the attachment of substrates.

Review everything we've said about enzymes. Then fill in the blanks and boxes appropriately.

- Enzymes promote chemical reactions by bringing _____ together in order that they may form products.

- The fact that enzymes interact with reactants by physically fitting together has given rise to the phrase _____ and _____ theory.

- Enzymes are known as organic _____.

- The physical location at which a substrate attaches to an enzyme is called the enzyme's _____ _____.

- When an enzyme has catalyzed a chemical reaction and the products are formed, the *enzyme itself* [❑ **is**   ❑ **is not**] consumed and is [❑ **unavailable**   ❑ **available**] to catalyze additional reactions.

ENZYME SPECIFICITY

The phrase enzyme specificity means that any given enzyme is designed so that it will hook up with only a *particular* set of substrates (reactants). In our first example, molecule E (above) is *specific* to reactants X and Y. The lock-and-key fit is special for E, X, and Y. Molecule E catalyzes only the reaction between X and Y to form Z.

Here's an enzyme called F.

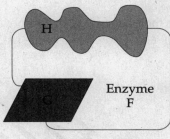

Molecule F

Molecule *F* is specific to substrates G and H. It catalyzes a reaction in which G and H get together to form some product. Enzyme specificity just means that one particular enzyme is designed for one particular set of substrates (reactants) and hence for one particular reaction. That's all.

Fill in the blanks.

- The fact that a given enzyme is designed to catalyze only a particular chemical reaction is described by the phrase _____ _____.

NEVER FORGET WHAT ENZYMES ARE MADE OF: *PROTEINS*

Enzymes are made of protein. Don't ever forget that: *All enzymes are protein . . . all enzymes are protein . . . all enzymes are protein.*

Know why enzymes are important. Start by thinking of a cell as a whole bucket of chemicals. If a whole bunch of chemicals are sitting around in a cell, what determines which ones will react with each other? ENZYMES DO. Enzymes decide which chemicals are going to get together and react. That's why they're important. Enzymes decide what chemical reactions a cell will conduct.

Sometimes enzymes need help in order to do their job. That's where *coenzymes* come into the picture. Coenzymes help enzymes work faster. Some enzymes can't work at all without a coenzyme. What do coenzymes have to do with you and the SAT II? Mainly this: *Vitamins act as coenzymes.* Memorize that statement for the SAT II, now that you know what it means.

TIME TO TALK ABOUT CELLULAR RESPIRATION

If you never took chemistry or physics, then you probably don't know what "energy" really means. That's okay. Just remember this: In order to do the things they do, cells *need* energy. They get that energy from a molecule called adenosine triphosphate (ATP). Here's what you have to remember about that.

A molecule of ATP, as the name indicates, is made of adenosine bonded to three molecules of phosphate. *There's a lot of energy in the bond that holds the third phosphate to the molecule.*

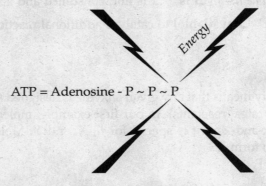

When a cell needs energy, it takes a molecule of ATP and removes the third phosphate. That releases energy. What's left over? A molecule of adenosine diphosphate (ADP) and one molecule of phosphate:

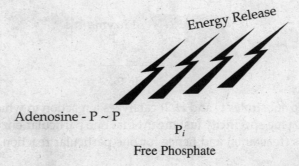

WHERE THE CELL GETS ITS ATP: GLYCOLYSIS, KREBS CYCLE, ELECTRON TRANSPORT CHAIN, AND OXIDATIVE PHOSPHORYLATION

We talked before about glycogen and said it's made from a large chain of glucose molecules. Glucose contains energy. Since glycogen stores glucose, its purpose is to *store energy*. When it comes to energy, think of it like this:

- ATP is like cash.
- Glucose is like a check.
- Glycogen is like a bank.

When a cell needs ATP, it goes to its glycogen stores and takes out a glucose molecule (a check). Then it cashes the glucose molecule in order to get ATP (cash). Here's how the check is cashed.

1. GLYCOLYSIS:

One glucose molecule yields two molecules of pyruvic acid. During that process, glucose releases some energy and some ATP is formed from ADP and phosphate.

$$\text{Glucose} \longrightarrow \text{2 Pyruvic acid}$$

and

$$\text{ADP} + P_i \longrightarrow \text{ATP}$$

Glycolysis occurs without oxygen. And when a process happens without oxygen we say it's <u>an</u>aerobic. Glycolysis is an <u>an</u>aerobic process.

Review what we've said about glycolysis. Then fill in the blanks and boxes appropriately.

- Glycogen serves as a bank that stores _____.

- Glycolysis takes energy from a molecule of _____ and uses it to produce some ATP from _____ and _____.

- The process of glycolysis produces ATP and converts one molecule of _____ to two molecules of _____ acid.

- The process of glycolysis [❏ **does**   ❏ **does not**] require oxygen.

2. THE KREBS CYCLE:

After glycolysis is complete, a process called the Krebs cycle takes place. When you think of the Krebs cycle, it's important to remember this: Since glycolysis produces *two* molecules of pyruvic acid from one molecule of glucose, there's a Krebs cycle for each of the pyruvic acid molecules. *One glycolysis reaction causes two rotations of the Krebs cycle*—one for each pyruvic acid molecule.

Here's what happens in the Krebs cycle. Energy is removed from the pyruvic acid molecule. Pyruvic acid is first converted to acetyl CoA. That produces more ATP molecules (from phosphate and ADP). The Krebs cycle also yields—THIS IS IMPORTANT—two substances called NADH and $FADH_2$.

Here's a picture of the Krebs cycle. *Don't memorize it.* But if the Biology Subject Test shows you a picture that looks like this, you'll know it's the Krebs cycle.

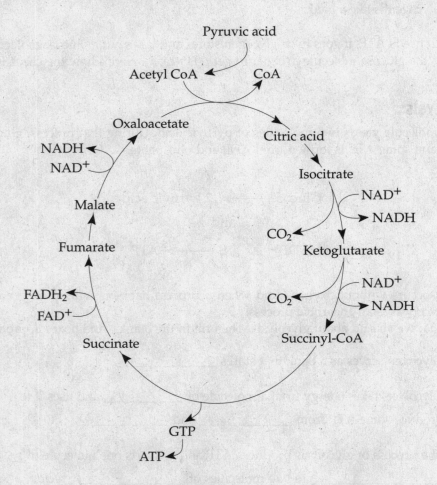

Here's something else that's very important about the Krebs cycle: IT REQUIRES OXYGEN. So, the Krebs cycle is very definitely an <u>a</u>erobic process.

Go back two paragraphs and review the Krebs cycle. Then fill in the blanks and boxes appropriately.

- The principal substance that enters the Krebs cycle is called _____ _____.

- The Krebs cycle consumes _____ _____ and produces some _____.

- The Krebs cycle [❏ does ❏ does not] require oxygen.

- The Krebs cycle occurs [❏ before ❏ after] glycolysis occurs.

- The Krebs cycle leaves the cell with CO_2, NA___ and FA___.

3. Electron Transport and Oxidative Phosphorylation:

When the Krebs cycle is complete, the cell undergoes a thing called electron transport. Electron transport is complicated and no one expects you to understand it fully. The Biology Subject Test does expect you to know a little about it. That's easy to do.

As we just said, the Krebs cycle leaves the cell with some additional ATP and—very important—NADH and $FADH_2$. The electron transport chain is a process in which NADH and $FADH_2$ hand down electrons to a chain of carrier molecules. First they go to one carrier, which hands them to the next, and the next, and the next, and finally to oxygen, which forms water and CO_2.

Don't worry about what this really means. When you read "electron transport chain," think: carrier molecules take electrons from NADH and $FADH_2$, handing them down from one to the next, and ultimately to oxygen to make water.

As the electrons are handed down, lots of energy is released. The cell uses that energy to make more ATP. The process by which the cell makes ATP from these cascading electrons is called oxidative phosphorylation.

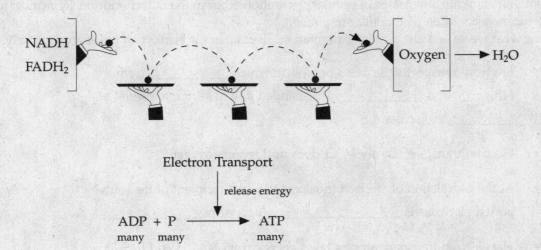

Electron transport produces lots of energy and this allows oxidative phosphorylation to produce lots of ATP—much more than glycolysis and the Krebs cycle produce.

You should know that electron transport REQUIRES OXYGEN: it's aerobic, just like the Krebs cycle. Here's something else to know. The carrier molecules we mentioned earlier contain iron. This helps explain why it's so important to have your daily dose of the stuff: Without iron, your electron transport chain can't function properly!

WE'D BETTER REVIEW

When glycolysis, the Krebs cycle, electron transport, and oxidative phosphorylation have all finished their jobs, the cell is left with a whole lot of ATP. That's the cell's energy. Where did this energy come from to begin with? It came from glucose that was stored in glycogen molecules. Glucose was removed from glycogen. Then the cell conducted:

- **Glycolysis:** Glucose was converted to two molecules of pyruvic acid and some ATP was formed along the way.

- **The Krebs cycle:** Each molecule of pyruvic acid was converted to acetyl CoA and entered the Krebs cycle, producing additional ATP, CO_2, NADH, and $FADH_2$.

- **Electron transport and oxidative phosphorylation:** NADH and $FADH_2$ gave up electrons to the electron transport chain and the electrons were handed down through a series of carrier molecules and then, finally, to oxygen to form water. As the electrons were passed along the carrier molecules, energy was released. That energy was immediately used to create ATP from ADP and phosphate during oxidative phosphorylation.

WHERE IN THE CELL DOES ALL OF THIS HAPPEN?

Everything we've just discussed—glycolysis, the Krebs cycle, electron transport, and oxidative phosphorylation—are, all together, called *cellular respiration*. Glycolysis occurs somewhere in the cytoplasm. But all of the other steps of cellular respiration occur in the mitochondrion. Remember to associate the mitochondrion with cellular respiration.

Review what we've said about cellular respiration. Then fill in the blanks and boxes appropriately.

- Electron transport describes a process in which _____ from NA_____ and _____ are "handed down" through a series of _____ molecules.

- Electron transport [❏ **does**   ❏ **does not**] require oxygen.

- At the completion of electron transport the final recipient of the transported electrons is _____, which forms _____.

- The electron transport process takes energy from NADH and $FADH_2$, using it to form ____ P through a process called _____ phosphor_____.

- The electron transport carrier molecules require _____, and an inadequate supply of _____ may cause the process to operate improperly.

- The cellular organelle most closely associated with cellular respiration is the _____.

Sorry, One Exception

Everything we just said about cellular respiration—glycolysis, the Krebs cycle, electron transport, and oxidative phosphorylation—is true if we're talking about aerobic organisms. But since the Krebs cycle and electron transport require oxygen, those processes CAN'T occur in anaerobic organisms—organisms that don't use oxygen.

So what about anaerobic organisms? How do they get ATP from glucose? In anaerobic organisms, the whole process of cellular respiration ends after glycolysis, which is an anaerobic process. Glycolysis doesn't require oxygen. Anaerobic organisms get energy from glucose and form ATP, but they only get as much as comes from the process of glycolysis. They can't get the energy out of the pyruvic acid that shows up when glycolysis is complete. They never participate in the Krebs cycle or electron transport and oxidative phosphorylation. Anaerobic respiration produces a lot less ATP than does aerobic respiration. That's important to remember.

In anaerobic respiration glycolysis *does* produce ATP and two molecules of pyruvic acid. But the pyruvic acid doesn't go into the Krebs cycle. Instead, it goes into a process called *fermentation*. In some organisms, fermentation produces something called lactic acid, and in other organisms—like yeast—it produces in some organisms ethyl alcohol (ethanol) instead.

So remember: When we're talking about aerobic organisms, cellular respiration means making lots of ATP from glucose through glycolysis, the Krebs cycle, electron transport, and oxidative phosphorylation. When we're talking about anaerobic organisms, cellular respiration means getting *some* (not a lot of) ATP from glucose through glycolysis only. Then it means converting the pyruvic acid to lactic acid or ethyl alcohol by fermentation.

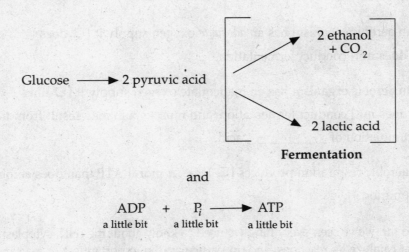

Anaerobic Respiration

AS A MATTER OF FACT, FERMENTATION CAN SOMETIMES OCCUR IN AEROBIC ORGANISMS

Think about heavy-duty exercise. It can make your muscles ache. Here's why. When your muscle cells exercise, they use a lot of energy. They take loads of glucose molecules and send them through the glycolysis process. Then, having produced some ATP and pyruvic acid, they try to do the Krebs cycle, electron transport, and oxidative phosphorylation.

But. . .

Suppose your muscles are working so hard and so fast that the blood stream can't deliver enough oxygen to keep the aerobic processes going. What's going to happen? The Krebs cycle and electron transport chain can't keep up. Even though your muscle belongs to an aerobic organism (you), it will act like it's <u>an</u>aerobic. The pyruvic acid that's formed from glycolysis will undergo fermentation and produce lactic acid. *The lactic acid causes muscle fatigue*.

If you notice that your legs ache while you're climbing ten flights of stairs, you'll know why. Your legs are using so much energy there isn't enough oxygen to permit <u>ae</u>robic respiration. The muscle cells in your legs conduct glycolysis, but lots of the pyruvic acid they produce is NOT entering the Krebs cycle. It's being converted to lactic acid. The lactic acid causes your legs to ache.

Review our discussion of aerobic and anaerobic respiration. Then fill in the blanks and boxes appropriately.

- <u>An</u>aerobic organisms [❏ **do** ❏ **do not**] conduct the *Krebs cycle*.

- <u>An</u>aerobic organisms [❏ **do** ❏ **do not**] conduct *glycolysis*.

- In <u>an</u>aerobic organisms the process of glycolysis is followed by a process called _____.

- The process of fermentation produces _____ or eth_____ from the _____ acid that results from glycolysis.

- If an aerobic organism has an *adequate* oxygen supply it [❏ **does** ❏ **does not**] conduct fermentation.

- If an aerobic organism has an <u>in</u>adequate oxygen supply it [❏ **does** ❏ **does not**] conduct fermentation, and muscle pain may result from the accumulation of _____ acid.

- <u>An</u>aerobic respiration produces [❏ **less** ❏ **more**] ATP than does aerobic respiration.

Well, well. So far, we've seen everything we need to know about the cell's cytoplasm, its organelles, the enzymes that catalyze its reactions, and, in particular, the important set of reactions that make for cellular respiration. Now, let's move on to the headquarters of the cell's activity, the nucleus.

Look at the Chromosomes

It is in the nucleus that we find the *chromosomes*. Chromosomes are made of **d**eoxyribo**n**ucleic **a**cid, **(DNA)**, and protein. Even though all chromosomes contain DNA, it is *not* true that all chromosomes are alike. The DNA in two different chromosomes can be put together in a way that makes them different, just like the wood that builds two different houses. The wood can be arranged in different ways so that the houses are different even though they're both made of wood. Fill in the blanks and boxes appropriately.

- All chromosomes are made of _____ and _____.

- All chromosomes [❏ **are**   ❏ **are not**] identical.

How Does DNA Build Chromosomes That Are Different from One Another?

Good question. Let's make it very easy. Let's look at a chromosome piece by piece. Each chromosome is made of *two DNA strands*. The DNA strands are linked up like a ladder, then twisted into a spiral shape, known as the double helix. The double helix was first discovered in 1956 by two scientists named Watson and Crick.

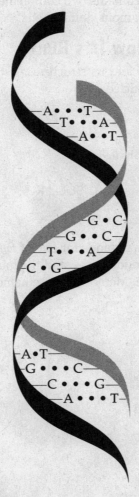

The two strands of DNA complement each other. When we think of a chromosome, we remember that it contains two complementary strands of DNA, put together in a shape that resembles a twisted ladder.

Now fill in the blanks appropriately.

- A DNA strand is described with the phrase _____ _____.

- Watson and Crick are famous for discovering the structure of _____.

- The two DNA strands found in a chromosome are said to be _____.

- The fact that double-stranded DNA forms a double helix was discovered by _____ and _____.

So, now we know a chromosome contains two complementary DNA strands put together like a ladder and then twisted to form the famous double helix.

ONE STRAND OF DNA AND HOW IT'S BUILT

Let's go into some detail about DNA and its structure. A strand of DNA is made of pieces called nucleotides. Here's a picture of one nucleotide.

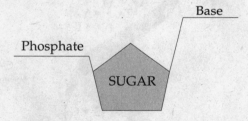

See where it says "base"? That word stands for one of *four* different things that can be sitting in that space. We call these things nucleotide bases, and the four nucleotide bases are: (1) adenine, (2) guanine, (3) cytosine, and (4) thymine. Memorize those names and fill in these blanks:

The four nucleotide bases are

1. _____,

2. _____,

3. _____, and

4. _____.

Because there are four different nucleotide bases, there are really four different kinds of nucleotides. Each type differs from the others in that it has one of the four nucleotide bases sitting in the space that says "base." Here are pictures of the four different kinds of nucleotides:

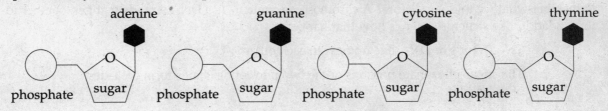

NUCLEOTIDES LINK UP IN A CHAIN AND MAKE A STRAND OF DNA

When many nucleotides link up, they form a chain and that chain is a strand of DNA. Since the four nucleotide types can link in all kinds of different orders, they can make all kinds of different DNA strands. For instance, look at these three strands.

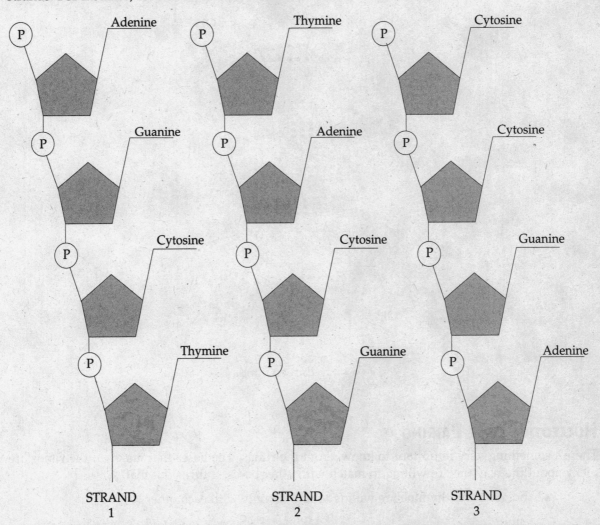

Each is a strand of DNA and each is made of nucleotides. But each strand *differs* from the others because of the *order* in which the nucleotides line up.

CRACKING CELLULAR AND MOLECULAR BIOLOGY ◆ 53

As You May Know, *Two* Strands Get Together To Make Double Stranded DNA

Remember what we told you before? A chromosome contains *two* DNA strands that get together to form a ladderlike structure. Here's how that happens:

1. Two strands of DNA line up next to each other.
2. The sugar-phosphate portions of the two nucleotide chains form the sides of the ladder.
3. The bases hook up to each other and form the ladder's rungs.
4. The ladder twists in a spiral and forms the double helix.

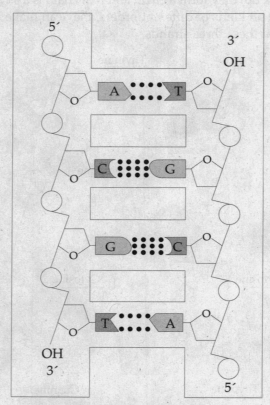

•••• = hydrogen bonds

Nucleotide Base Pairing

There's something very important to know about the rungs. The bases that make up the rungs are picky about the way they're willing to match with other bases. It turns out that:

- adenine and thymine are willing to link up only with each other

  and

- cytosine and guanine are willing to link up only with each other.

Adenine won't bond to cytosine or guanine. It bonds only with thymine. Thymine feels exactly the same way. It only likes adenine.

Cytosine won't bond to adenine or thymine. It bonds only with guanine. Guanine feels exactly the same way. It likes only cytosine.

Here's how to remember who likes whom. Make an alphabetical list of the bases.

Adenine   Cytosine   Guanine   Thymine

Then remember that:

- the two bases on the ends bond with each other

  and

- the two bases in the middle bond with each other.

Adenine likes thymine and cytosine likes guanine. Another way to say all of this is to say that adenine and thymine make a nucleotide base pair. Cytosine and guanine make the other nucleotide base pair.

Fill in the blanks.

- Guanine forms a base pair with _____.

- Thymine forms a base pair with _____.

- Cytosine forms a base pair with _____.

- Adenine forms a base pair with _____.

The Biology Subject Test might present you with a strand of DNA and tell you the order of its bases. Then it will ask *you* to figure out the order of bases on the *complementary* strand. Since you know about base pairing, that's easy to do.

Suppose the test writers tell you that a particular DNA strand has the sequence:

THYMINE – THYMINE – GUANINE – CYSTOSINE – ADENINE – GUANINE.

Then suppose you're asked to determine the base sequence on the *complementary* strand. You do that by knowing the base-pairing rules. Since adenine/thymine form one base pair, and guanine/cytosine form another, you can easily figure out that the complementary strand has the sequence:

ADENINE – ADENINE – CYTOSINE – GUANINE – THYMINE – CYTOSINE.

Here, try it yourself. Choose the correct letter:

- If a particular region of a chromosome has, on *one* of its strands, the base sequence adenine-cytosine-guanine, the complementary strand has the base sequence

    (A) guanine-guanine-thymine
    (B) adenine-cytosine-guanine
    (C) thymine-guanine-cytosine
    (D) cytosine-guanine-thymine

DNA REPLICATES ITSELF

One of the big deals about DNA is that it can replicate itself. The Biology Subject Test writers love to ask questions about that. Now that you know how DNA is put together, it's easy to understand how it replicates. Let's say we're dealing with a DNA molecule that's all assembled into its double helix. Okay. It's going to replicate in four simple steps.

1. The helix unwinds, and the two strands separate.

2. Alongside *each* separated strand, nucleotides line up and form a new *second* strand. When these nucleotides line up, they follow the base-pairing rules. Adenines match up with thymines, and thymines match up with adenines. Guanines match up with cytosines, and cytosines match up with guanines.

3. Bonds form between the base pairs and thus form new rungs. Bonds also form along the sugar-phosphate component of the newly aligned nucleotides so that the newly formed ladder has a new side too.

4. The new double-stranded molecule twists up into a double helix.

We start with one DNA molecule. The strands separate and *for each strand* the cell makes a new complementary strand. So we end up with *two* DNA molecules. The DNA molecule has replicated.

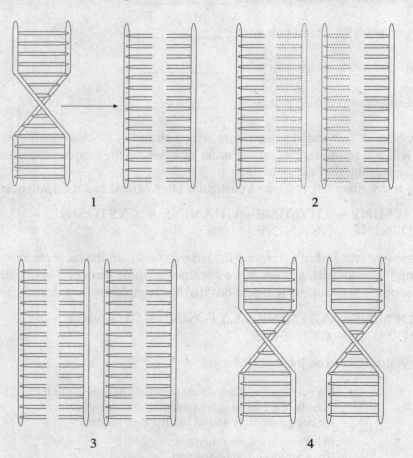

AN IMPORTANT BIOLOGY SUBJECT TEST WORD: TEMPLATE

Think of a replicating DNA molecule. Its strands separate. Now think about *one* separated strand. That strand causes the formation of a new complementary strand with nucleotides that are ordered according to the base-pairing rule. Another way of saying all of that is this:

"Each DNA strand acts as *template* for the formation of a new complementary strand."

When we say that one strand acts as template for the formation of another, we mean that its nucleotide bases direct the construction of a complementary strand that forms alongside it. That's all.

Let's make sure you've got all of this straight. Designate the order of events through which a DNA molecule replicates by writing the numbers 1, 2, 3, and 4 next to the appropriate statements.

___ Each strand serves as a template.

___ The double helix unwinds and its DNA strands separate.

___ The newly formed double-stranded molecule twists to form a double helix.

___ Nucleotides align themselves along the template according to the base-pairing rules.

CHROMOSOMES AND THE WHOLE ORGANISM: THE SAME SET IN EVERY CELL

You are an organism. You have cells, your cells have nuclei, and the nuclei contain chromosomes. If we're going to think about chromosomes, let's start by thinking about *you* and *your* chromosomes.

You're made up of billions of cells. Every non-sex, or somatic cell has a set of 46 chromosomes sitting in it. Now get this: *the set of chromosomes that's sitting in any one of your somatic cells is identical to the set that's sitting in every other of your somatic cells.* It's not like your skin cells have one set of chromosomes and your kidney cells have another. *All of your cells have exactly the same set of chromosomes in their nuclei.* The set itself consists of 46 chromosomes, but every somatic cell in your body has exactly the same *set*.

What goes for you goes for every organism in the whole world. Think about your best friend. Her (or his) chromosomes are different from yours, that's for sure. But the set of chromosomes in any one of *her* cells is absolutely identical to the set that's in every other one of *her* cells. Think about your dog, cat, turtle, or hamster. Think about some tree on your block. Think about any organism you like and this will always be true: Within any one organism the set of chromosomes that's sitting in one cell's nucleus is identical to the set that's sitting in every other.

Okay? Fill in the boxes appropriately.

- If two individuals are of the same species, then the chromosomes in one individual [❏ are ❏ are not] identical to the chromosomes of the other.

- If two cells are taken from the same individual, the chromosomes in one cell [❏ are ❏ are not] identical to the chromosomes in the other.

Sorry, Another Exception (Why Do We Keep Doing That?)

We just taught you that, within any individual, all cells have identical sets of chromosomes. That's true. We also told you that two different individuals do not have the same sets of chromosomes. That's *almost* always true. The exception is: identical twins. Each identical twin has exactly the same chromosomes as the other. That's why the twins are identical.

Chromosomes Come in Pairs: Homologous Chromosomes

We said that human beings have a total of 46 chromosomes. That's true. But these 46 chromosomes occur in *pairs*: human beings have *23 pairs of chromosomes*. Lots of other species also have their chromosomes situated in pairs. Different species have different numbers of chromosomes and hence different numbers of pairs. Forty-six happens to be the total number of chromosomes in *human* cells and hence human cells have 23 pairs of chromosomes. When we talk about a pair of chromosomes we say "homologous chromosomes." The phrase "homologous chromosomes" just means a pair of chromosomes.

Let's think about one pair of chromosomes and ask this question: Are the two members exactly alike? The answer is no. The two members of any particular pair of chromosomes are *similar*, but not *exactly* alike. They're like brothers or sisters who look alike and have a lot in common, but they're definitely not identical.

Fill in the blanks and boxes appropriately.

- The two members of a chromosome pair are called _____ chromosomes.

- Homologous chromosomes [❏ **are** ❏ **are not**] similar to each other.

- Homologous chromosomes [❏ **are** ❏ **are not**] identical.

Look at this pair of chromosomes: A and B. A and B are homologous.

A   B

Since A and B are homologous they have a lot in common, but they're not *exactly* alike.

Now you know what chromosomes are and you know they occur in homologous pairs. You also know that they're made of DNA and protein and that they replicate. That leaves us with one little question:

What do chromosomes *do*?

Chromosomes contain genes. Genes make an organism what it is, at the moment it's first conceived. They do that, basically, *by determining what proteins its cells will manufacture.* You know that some proteins serve as enzymes. By deciding what enzymes will be hanging around in the cells, genes determine *what chemical reactions will and will not occur* in the cells. *That's* what makes an organism, more or less, what it is.

Fill in the blanks and boxes appropriately.

- An organism's genes are contained in its _____.

- The chromosomes exert their influence over the organism by directing the production of [❑ **carbohydrates** ❑ **proteins**].

- Control over the organism's synthesis of proteins is particularly significant because some proteins serve as _____ which, in turn, determine the _____ reactions that the cell will conduct.

How Chromosomes Govern Protein Synthesis: Transcription and Translation

You know that chromosomes themselves are made of deoxyribonucleic acid (DNA). Well, there's another chemical similar, but not identical, to DNA, and it's called ribonucleic acid (RNA). Chromosomes are not made of RNA. They're made of DNA. But chromosomes can cause the production of RNA. RNA, in turn, causes the production of protein. We'll tell you how it all works, but first we'll tell you about RNA and how it's different from DNA.

1. RNA is single-stranded. That makes it different from DNA. DNA is double-stranded.

2. RNA, like DNA, consists of a sugar-phosphate backbone.

But...

1. In RNA the sugar is called <u>ribose</u>, not deoxyribose. That's why the full name for RNA is <u>ribo</u>nucleic acid.

2. In RNA, three of the bases are adenine, guanine, and cytosine, just as they are in DNA. But the fourth base is called <u>uracil</u>. So, in RNA, there's no thymine. There's uracil instead.

When we list alphabetically the bases that belong to <u>R</u>NA we get: adenine, cytosine, guanine, and <u>uracil</u> (*not* thymine).

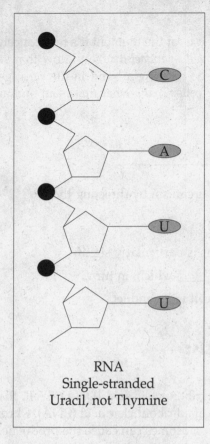
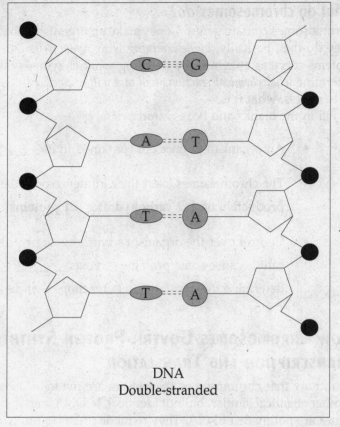

RNA
Single-stranded
Uracil, not Thymine

DNA
Double-stranded

Review our comparison of RNA and DNA. Then fill in the blanks and boxes appropriately.

- RNA nucleotides [❏ do ❏ do not] contain the exact same bases as do DNA nucleotides.

- RNA [❏ is ❏ is not] a double-stranded molecule.

- DNA [❏ is ❏ is not] a double-stranded molecule.

WHERE RNA COMES FROM

Remember DNA replication? The DNA molecule unzips itself and each strand then serves as a template for a new complementary strand. A new ladder is formed with rungs that are really bonds between base pairs. Adenine bonds to thymine and cytosine bonds to guanine, just as you saw above. *Sometimes*, however, a DNA molecule unzips when it's not planning to replicate. Rather, it's planning to create a molecule of RNA. The whole thing is really pretty simple.

The DNA unzips in the usual way. Two strands are exposed. One of the strands causes RNA nucleotides to line up next to it as though they were going to form a new ladder. Right next to a DNA strand, RNA nucleotides line up. The lineup is pretty much the same as the one that we see when DNA replicates, EXCEPT that thymine doesn't line up because it's not one of the RNA nucleotides; uracil takes its place.

A cytosine lines up next to each guanine on the DNA strand and a uracil lines up next to each adenine on the DNA strand. The newly lined up RNA nucleotides form bonds along the sugar-phosphate backbone. Before you know it, an RNA molecule is born.

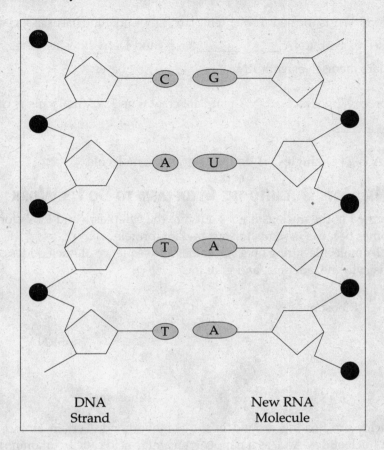

DNA Strand   New RNA Molecule

The two strands do **NOT** then bond together crosswise like a ladder. They stay apart. Remember: RNA is a *single*-stranded molecule. The DNA strand goes back to join its complementary strand and the RNA strand is let loose.

One more thing: There's actually more than one kind of RNA in this world. The kind of RNA we've just discussed is called messenger RNA, or mRNA.

We just told you how DNA unwinds and how one of its strands serves as a template for the production of a messenger RNA (mRNA) molecule. Now let's tell you the word for that process. The process is called *transcription*. Transcription is complete when the new RNA has copied the sequence of nucleotide bases directly from the exposed DNA strand.

Review what we've said about the formation of mRNA molecules. Then fill in the blanks and boxes appropriately.

- The creation of an mRNA strand requires first that a strand of _____ separate from its complementary strand.

- mRNA [❏ **is** ❏ **is not**] formed from a *template* that arises from a D̲NA molecule.

- mRNA [❏ is ❏ is not] formed from a *template* that arises from a RNA molecule.

- When mRNA is formed from a template, DNA adenine bases associate themselves with RNA _____ bases, and DNA cytosine bases associate themselves with RNA _____ bases.

- The process by which a DNA template creates an RNA molecule is called _____.

- mRNA is a [❏ single- ❏ double-] stranded molecule.

MESSENGER RNA GOES OUT INTO THE CYTOPLASM TO DO ITS WORK

Once mRNA is formed in the nucleus it goes out into the cytoplasm and looks for some ribosomes. It looks for ribosomes because it wants to start making protein.

Here's an mRNA molecule sitting on a ribosome. The sequence of nucleotides on a section of this mRNA molecule is cytosine, adenine, and guanine.

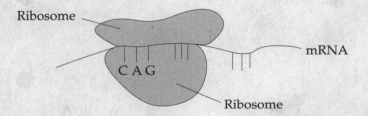

Some amino acids are floating around in the neighborhood. Certain amino acids like certain sequences of RNA nucleotides. And, as it happens, an amino acid called glutamine likes the sequence cytosine, adenine, guanine. When glutamine sees the sequence cytosine, adenine, guanine, it steps up to the ribosome and gets close to it.

Look at the next two sequences of nucleotides on the mRNA molecule. One is uracil, uracil, guanine, and the next is cytosine, adenine, uracil. Do some amino acids "like" these sequences? Yes. It just so happens that the amino acid leucine likes UUG and the amino acid histidine likes CAU. Leucine gets close to UUG, and histidine gets close to CAU.

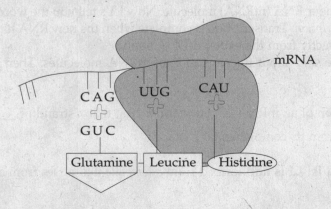

WHAT WE MEAN WHEN WE SAY "LIKE": CODONS

What do we mean when we say certain amino acids like certain sequences of mRNA nucleotides? We mean that certain amino acids tend to associate themselves with certain sequences of mRNA.

We call each sequence a "codon" or "triplet." A codon is a series of three mRNA nucleotides that somehow attracts or codes for a particular amino acid. Don't worry about *how* the codon attracts—it just does. CAG, UUG, and CAU are all codons. The CAG codon codes for glutamine. The UUG codon codes for leucine and the CAU codon codes for histidine.

Look what the codons have done. They've made glutamine, leucine, and histidine line up in a row. These three amino acids *were* floating around loose and lonely in the cytosol. But when their codons showed up on the ribosomes they went over and parked themselves nearby. They parked in the order glutamine-leucine-histidine, because that's the way their codons were ordered. Those codons, in turn, are part of the mRNA molecule that a DNA template formed in the nucleus.

Notice, then, that the ordering of the amino acids on the ribosome was actually caused by the ordering of DNA nucleotides in the chromosomes.

- The DNA molecule ordered the mRNA nucleotides

  and

- The mRNA molecule ordered the amino acids.

When amino acids find their codons and line up on the ribosome, we call this process "translation." Remember to associate

1. *transcription* with the formation of mRNA from DNA

   and

2. *translation* with amino acids that find their codons and line up on a ribosome.

LET'S REVIEW: HOW ARE PROTEINS MADE?

Messenger RNA leaves the nucleus, goes out into the cytoplasm, and finds a ribosome. As more and more amino acids are added they link up. When a whole chain is lined up, **peptide bonds** form between adjacent amino acids. A chain of amino acids linked together by peptide bonds is a polypeptide, otherwise known as a protein.

There's one more thing we should mention. Amino acids that float around in the cytoplasm have a special way of getting to the ribosomes. A molecule we call transfer RNA or tRNA takes them. You don't really have to know much about tRNA. Just remember that it's a taxicab that takes amino acids to the ribosome so they can meet up with their codons.

Review everything we've said about protein synthesis from the time mRNA leaves the nucleus. Then fill in the blanks and boxes appropriately.

- mRNA leaves the nucleus after it is formed and locates itself on a

  _____, which is the site of protein synthesis.

- A series of three mRNA nucleotides that codes for a particular amino acid

  is called a _____.

- For every amino acid, there [❑ **is** ❑ is not] a codon.

- Amino acids are brought to the ribosome by a substance called [❑ messenger ❑ **transfer**] RNA.

- Once a series of amino acids has aligned itself along a ribosome in accordance with the associated sequence of mRNA codons, _____ bonds form between adjacent amino acids to form a _____.

There's one more kind of RNA. It's called ribosomal RNA or rRNA and it's not too important. You just need to know that the ribosome itself is, in part, actually *made of* RNA. *That* RNA is called rRNA. So there are three kinds of RNA:

1. mRNA, which is made in the nucleus by transcription from a DNA template,

2. tRNA, which brings amino acids to the ribosome so they can meet up with their mRNA codons,

   and

3. rRNA, which is what ribosomes are, in part, actually made of.

HOW A WHOLE *CELL* REPRODUCES ITSELF: MITOSIS

You need to know about mitosis. That's the process by which cells divide. Before we talk about mitosis itself, we'll tell you what happens *before* mitosis occurs. To make things easy, we'll talk about a human cell.

Most human cells have 46 chromosomes in their nuclei. The chromosomes hang out in pairs, so we can say that nuclei have 23 pairs of homologous chromosomes. Before a cell undergoes mitosis, every single chromosome in its nucleus replicates. In a human cell, all 46 chromosomes have to replicate. We've already discussed how that's done.

INTERPHASE

Interphase is the time during which chromosomes replicate, but a lot of other things happen during interphase. The cell carries out all of its normal activities during interphase. Interphase is sometimes called the resting stage of the cell—not because the cell is taking it easy—but because the cell is not actively dividing.

Once interphase is over, the cell has replicated every one of its 46 chromosomes. How many chromosomes does it have? The answer would *seem* to be 46 × 2 = 92. When a cell has finished interphase you'd think it has 92 chromosomes and basically, you're right. But biologists start playing games with words at this point and you have to watch out.

WATCH OUT FOR WHAT? THIS WORD: *CHROMATID*

After interphase, *each chromosome and the duplicate it just made* are tied together at their center by a spot called a **centromere.** The two chromosomes and the centromere make *one* united physical structure.

Biologists look at the whole structure—the two chromosomes joined by a centromere—and call *it* a chromosome. Then they use the word chromatid to describe each of the chromosomes.

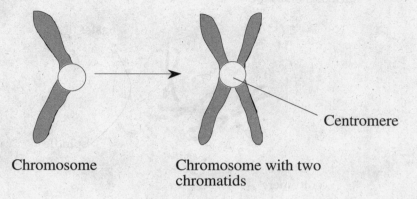

Chromosome   Chromosome with two
                  chromatids

When interphase is over, all of the cell's 46 chromosomes have doubled. We might *want* biologists to say the cell has 92 chromosomes, but they don't say that. Instead they say the cell *still* has 46 chromosomes, each now consisting of two chrom<u>atids</u>. It's a little word game they play.

Let's make sure you've got it. Fill in the blanks and boxes appropriately.

- During a stage called interphase [❑ **all**   ❑ **only some**] of the cell's chromosomes replicate.

- According to modern biological terminology, a human cell, after interphase, has a total of [❑ **92**   ❑ **46**] chromosomes, each chromosome having at its center a _____ that joins the chromatids together.

- A human cell, after interphase, has a total of _____ chromosomes, each made up of two _____.

- DNA replication is [❑ **the only**   ❑ **one of many**] process(es) that takes place during interphase.

After interphase, mitosis begins.

MITOSIS HAPPENS IN FOUR STEPS

Step 1 is called *Prophase*. The centrioles move away from each other to opposite sides of the cell. They also form a bunch of fibers called the mitotic spindle. The chromosomes thicken and we can see them (under the microscope, of course). The nuclear membrane begins to break up, too.

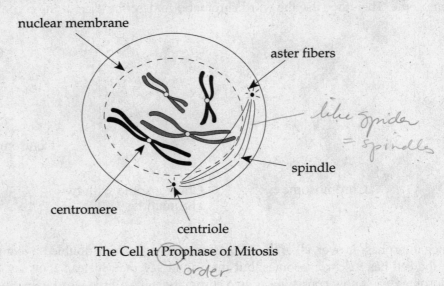

The Cell at Prophase of Mitosis

Step 2 is called *Metaphase*. The chromosomes line up on the spindle fibers at the equator of the cell. One chromosome (which really means a chromosome and its duplicate—two chromatids) lines up on each spindle fiber.

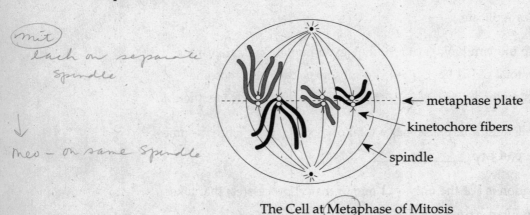

The Cell at Metaphase of Mitosis

Step 3 is called *Anaphase*. The centromere that joins each pair of chromatids splits in two so that each chromatid separates from its pair. And guess what? Now each chromatid is once again called a chromo*some*. So once the centromeres split, even a biologist will admit that the cell briefly has 92 chromosomes. The newly separated chromosomes move toward opposite poles of the cell with the help of the spindle fibers.

Anaphase:
back, again,
anew

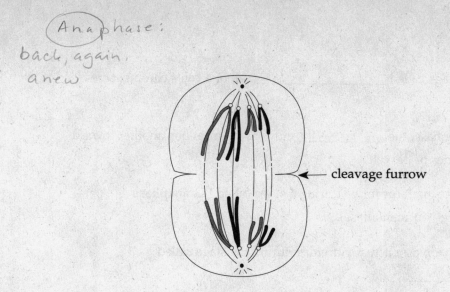

The Cell at Anaphase of Mitosis

Step 4 is called *Telophase*. A nuclear membrane forms in each new cell and two daughter cells result, each of which has 46 chromosomes. The cytoplasm then divides, during a process called cytokinesis.

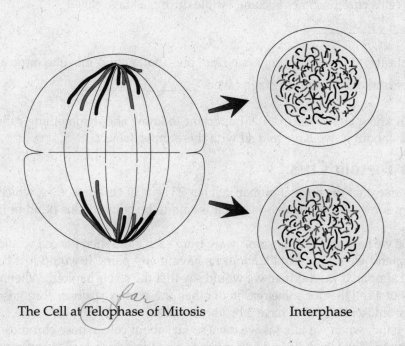

The Cell at Telophase of Mitosis   Interphase

Then, of course, the daughter cells enter interphase. So let's get the order straight:

- Before mitosis: Interphase

    Mitosis: (1) prophase (2) metaphase (3) anaphase
    (4) telophase

Review the events associated with each phase of mitosis. Then fill in the blanks and boxes appropriately.

- During a stage called _____, all of a cell's chromosomes replicate.

- During prophase, the _____ move away from one another toward opposite sides of the cell.

- The spindle apparatus forms during a stage called [❏ **anaphase** ❏ **prophase** ❏ **metaphase**].

- The cell's cytoplasm actually divides during the stage called _____.

- The centromere divides during a stage called _____.

- Cytokinesis involves the division of [❏ **chromosomes** ❏ **cytoplasm**].

- The cell's chromosomes become visible during a stage called _____.

- Duplicate chromosomes separate from one another and move to opposite poles of the cell during a stage called _____.

Now that you know about mitosis, it's important to know about something called meiosis. But before we tell you about it, we have to deal with this simple subject.

HAPLOID AND DIPLOID CELLS

You know, of course, that the normal human cell has 23 *pairs* of chromosomes, which means it has a total of 46 chromosomes. Each member of a pair is said to be homologous (kind of like a sibling) to the other.

Now, suppose we took one chromosome away from each pair. Then the cell would have a total of 23 chromosomes, and we wouldn't talk about its having any pairs. It would just have 23 chromosomes, plain and simple. If we did that, we would say that the cell is haploid. When we say haploid, we're talking about a cell that for some reason or other doesn't have *pairs* of chromosomes. It's a cell in which each chromosome doesn't have a brother or sister.

On the other hand, when we talk (as we have so far) about cells whose chromosomes *do* exist in pairs, we say diploid. The word diploid just refers to a cell in which each chromosome does have a sibling: it has chromosome pairs. The ordinary human cell with 23 pairs of chromosomes is diploid. If we stole one member of each pair, it would be haploid.

The words "diploid number" mean the number of chromosomes a cell has when it's in the diploid state. For a human cell the diploid number is 46. "Haploid number" means the number of chromosomes a cell has when it's in the haploid state. Naturally, the haploid number is always one-half the diploid number. For a human being the haploid number is 23. Don't go crazy trying to *understand* all this stuff. Just know it.

Review what we've said about haploid and diploid cells. Then fill in the blanks and boxes appropriately.

- The word [❑ haploid   ❑ diploid] refers to a cell for which each chromosome has a homologous chromosome to form a chromosome pair.

- The word [❑ haploid   ❑ diploid] refers to a cell for which each chromosome does NOT have a homologous chromosome to form a chromosome pair.

- If, for a particular organism, the diploid number of chromosomes is 10, the haploid number is _____.

We've told you about the words haploid and diploid because that will make it easier to explain the next subject: meiosis.

THE FORMATION OF GAMETES: MEIOSIS

When a man and woman decide to have a baby, both will contribute chromosomes. The trouble is that each parent has 46 chromosomes. We need a way in which each parent gives the baby only 23 chromosomes so we get a baby that has only 46 chromosomes, not 92. That's what meiosis is all about.

When we say meiosis, we're talking about the formation of sperm cells and ova. Sperm cells and ova (egg cells) are called gametes. They're the cells that get together and form a new organism's first cell. That first cell is called a zygote. So two gametes combine to form a zygote.

Sperm and ova are the only human cells that are haploid. Each has 23 chromosomes. There's no pairing—no siblings. When an ova and sperm get together, each chromosome from the ovum pairs up with a chromosome from the sperm. The newly formed cell—the zygote—is diploid. The diploid zygote begins the new organism's development.

MEIOSIS AND THE FORMATION OF SPERM CELLS: SPERMATOGENESIS

Since ova and sperm are the gametes, the formation of ova and sperm is called gametogenesis. When we talk about formation of sperm cells, we call it spermatogenesis. Spermatogenesis requires meiosis (not mitosis). Here's how it works.

We start with an ordinary diploid cell called a spermatogonium. (It's gonna become a sperm cell.) Spermatogonia live in a place called the seminiferous tubule. So where are seminiferous tubules located? In the testes. The testes are the male gonads. One spermatogonium, which is diploid, undergoes meiosis and produces four sperm cells, which are haploid. Here's how it happens.

The spermatogonium cell undergoes DNA replication during interphase just as it would if it were about to conduct ordinary mitosis. All the chromosomes replicate, and we're left with a cell that still has 46 chromosomes, each made up of two chromatids joined by a centromere.

Then come the four stages of meiosis, named after the stages of mitosis: prophase, metaphase, anaphase, and telophase. The big difference between meiosis and mitosis is synapsis—the homologous chromosomes lie next to each other. Remember the 23 pairs of chromosomes in humans? Well in meiosis, they come together to form a structure called a tetrad-four chromatids. If pieces of chromosomes are exchanged this is called crossing over. Now let's look at metaphase.

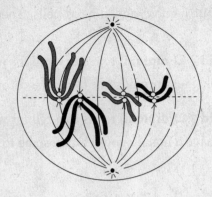

Mitosis

Metaphase: In *mitosis*, metaphase means that each of the 46 chromosomes line up on a separate spindle fiber. In meiosis that doesn't happen. Instead, each chromosome, *together with its homologous partner*, lines up on a spindle fiber.

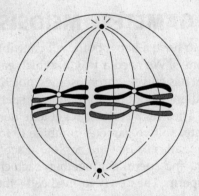

Meiosis

Anaphase: Then, in anaphase, centromeres don't divide. Instead, the homologous *pairs* separate. They say goodbye to each other: brother leaves brother, sister leaves sister. They move to opposite ends of the cell. The cell then finishes dividing and we're left with a strange situation. We have two cells, each with 23 chromosomes. Every chromosome is made of two identical chromatids, joined by a centromere. Each cell DOES NOT have 23 sibling-type pairs of chromosomes. Rather, it has 23 chromosomes, each of which is made of two *identical* chromatids.

Both cells get ready to divide again. This time they do it pretty much the way it's done in mitosis. Each chromosome lines up on a spindle and the centromeres DO split. That means the chroma*tids* are called chromo*somes* once again.

When the second division is over, we have four cells. Each has 23 chromosomes and those 23 chromosomes have no siblings. (They lost their sibling at the beginning of anaphase when chromosome pairs lined up on spindles and each member of the pair separated from the other.) They're haploid cells. We started with one diploid cell—the spermatogonium and finished with four haploid cells—the sperm cells, also called spermatozoa.

Let's review. We're going from one diploid spermatogonium to four haploid spermatozoa.

1. The spermatogonium replicates all of its chromosomes during interphase. It's left with 46 chromosomes, but each has two chroma<u>tids</u>, joined by a centromere.

2. The cell undergoes prophase just as it does in mitosis.

3. The cell undergoes metaphase, but it's different from mitosis. *Pairs* of chromosomes line up on spindles. So we see *two* centromeres on each spindle fiber.

4. The cell undergoes anaphase, but the centromeres don't split. Instead, the chromosome *pairs* separate. The cell finishes dividing, and we're left with two cells. Each cell has 23 chromosomes, and each chromosome is made up of two chromatids, still joined by a centromere.

5. Each of these two cells then goes through *another* metaphase, prophase and anaphase, which do resemble mitosis. Centromeres line up along spindle fibers, one centromere per fiber. The centromeres DO divide.

6. At telophase the cells finish dividing and since steps (5) and (6) happen to *two* cells that resulted from (4), we end up with four cells, each of which has 23 chromosomes. The 23 chromosomes don't have homologous partners: they're haploid.

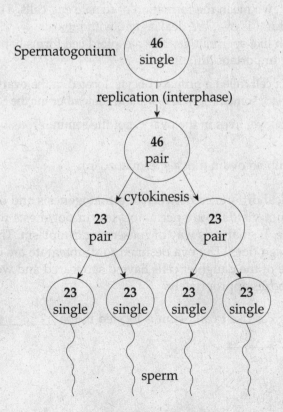

Spermatogenesis

Review it, master it, and fill in the blanks and boxes appropriately.

- Prophase of meiosis [❏ is ❏ is not] similar to prophase of mitosis.

- The first metaphase of meiosis differs from metaphase of mitosis in that a _____ of chromosomes line up on each spindle fiber.

- The first anaphase of meiosis differs from anaphase of mitosis in that centromeres [❏ do ❏ do not] divide.

- During the second metaphase, anaphase, and telophase of meiosis, _____ divide, cells divide, and the resulting four daughter cells contain _____ chromosomes, each of which [❏ does ❏ does not] have a homologous partner.

- The four sperm cells that result from spermatogenesis are _____loid.

Now that you know how spermatozoa are formed, let's talk about the formation of ova.

THE FORMATION OF OVA: OOGENESIS

When we say "oogenesis," we mean the formation of female egg cells. The fancier name for an egg cell is an ovum and the plural is ova. We deal, again, with meiosis.

Oogenesis is very much like spermatogenesis: a diploid cell forms haploid cells through meiosis. There are, however, some important differences.

1. We start with a cell called a primary oocyte, located in the ovary. The ovary is the female gonad, something you'll want to remember for the SAT II.

2. The primary oocyte lives in the ovary (not the seminiferous tubules of the testes).

3. We end up with an ovum (not a spermatozoan).

There's one more critical difference between spermatogenesis and oogenesis. In spermatogenesis, one spermatogonium yields four spermatocytes. In oogenesis we also get four cells, but three of them disintegrate. It's nature's way of conserving cytoplasm. The sperm cell doesn't need much cytoplasm, but the egg does. The ova destined to disintegrate are called polar bodies. When oogenesis is all over, three of the daughter cells have disappeared and we have only one ovum.

Fill in the blanks and boxes appropriately.

- The cell that gives rise to an ovum is located in the _____ and is called a _____.

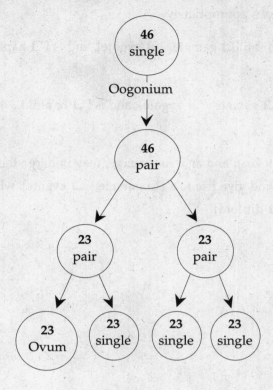

Oogenesis

- Polar bodies are those cells that [❏ **disintegrate during oogenesis** ❏ **remain after other ova have disintegrated**].

- Oogenesis ultimately gives rise to [❏ **one ovum** ❏ **four ova**] with the [❏ **haploid** ❏ **diploid**] number of chromosomes.

- Gametogenesis includes _____ and _____, and produces _____ and _____ , respectively.

Two Haploids Make a Diploid: Fertilization and the Zygote

When a spermatozoan and an ovum meet, they form a diploid cell. That diploid cell is a zygote. When sperm and ovum merge we call it "fertilization." The ovum has been fertilized. The two cells kind of join into one. Since each haploid cell had 23 chromosomes, the resulting zygote has *23 pairs* of homologous chromosomes.

The zygote divides by mitosis. It keeps dividing and dividing and dividing. All kinds of things happen to it and after nine months or so, we get a new person.

Fill in the blanks and boxes appropriately.

- A spermatozoan is a [☐ **gamete** ☐ **zygote**], and is [☐ **haploid** ☐ **diploid**].

- An ovum is a [☐ **gamete** ☐ **zygote**], and is [☐ **haploid** ☐ **diploid**].

- When a spermatozoan and an ovum merge, they undergo the process of _____, and give rise to a [☐ **gamete** ☐ **zygote**], which is [☐ **haploid** ☐ **diploid**].

Cracking Genetics

REMEMBER WHY CHROMOSOMES ARE IMPORTANT: THEY CONTAIN GENES

From chapter 1 you should remember that:

- chromosomes contain very long strands of DNA
- DNA is a chain of nucleotides
- a chain of DNA nucleotides can direct the production of a molecule of mRNA, which is also a chain of nucleotides
- mRNA travels from the nucleus to the cytoplasm and locates itself on a ribosome
- a series of three mRNA nucleotides is a "triplet" or "codon" which codes for a particular amino acid
- amino acids line up along the ribosome according to the order of the mRNA molecule's codons located there

- the amino acids link themselves together to form peptide bonds and a **polypeptide** is formed.

This process is known as protein synthesis.

Now that we've summarized (yet again) the relationship between chromosomes and the creation of proteins, we can answer this question:

WHAT, EXACTLY, IS A GENE?

These days when biologists say "gene," they're talking about the whole process we just described and, specifically, they're talking about the portion of a chromosome that gives rise to enough mRNA codons to make *one full protein molecule*. Sometimes that's called the one-gene-one-protein theory, which just means that when we say "gene," we're thinking of some portion of a chromosome that gives rise, ultimately, to *one protein molecule*. A gene is any part of any chromosome that is responsible for the creation of one protein molecule.

A chromosome is basically a rather long segment of DNA, and one chromosome contains many, many genes. It takes three nucleotides to make one mRNA codon, and even though one codon codes for only one amino acid and a single protein is a long, long chain of amino acids, a single chromosome is so *very, very long* that it may give rise to hundreds of proteins. That's hard to imagine, but it's true.

So remember: When we say "gene," we're talking about one portion of one chromosome. What portion? A portion that ultimately produces—via mRNA and the ribosomes—one protein. Remember, also, that one chromosome has enough nucleotides to bring about the production of loads of different proteins. This is another way of saying that one chromosome contains a large number of genes.

Okay? Fill in the blank lines and boxes appropriately:

- The term "gene" refers to that portion of a chromosome that ultimately produces one [❏ **amino acid** ❏ **protein**] via mRNA and the assembly process that occurs on the ribosome.

- A single gene is responsible, ultimately, for the production of a single _____.

- Proteins are most significant for the organism in that they function as _____, which catalyze biological reactions.

NOW PAY ATTENTION: WE'RE GOING TO TALK ABOUT GENETICS AND INHERITANCE

When you take the Biology Subject Test you'll definitely get some questions about genetics and inheritance. So we'll teach you what you have to know. You already know that:

- Genes direct the production of mRNA.
- mRNA contains three-nucleotide-long segments called codons.
- Each codon codes for a single amino acid.

- Amino acid chains become proteins.
- Proteins serve as enzymes.

The enzymes inside an organism's cells determine which chemical reactions will and will not occur, and *that* determines what the organism is like. So, an organism's genes determine its features, its characteristics, its appearance—in other words, its *traits*. They decide what traits the organism will possess.

When talking about an organism's traits, we say "phenotype." If you're a cat and your fur is gray, then we say that, in terms of fur color, your phenotype is gray. If your fur is white, then we say, in terms of fur color, your phenotype is white. If we meet a person who has diabetes, we say that, in terms of diabetes, his phenotype is diabetic. If we meet another person who does not have diabetes we say that in terms of diabetes, her phenotype is nondiabetic, or we might say that her phenotype is normal. So when we say "phenotype," we're just talking about the actual traits that an organism does and does not possess.

PHENOTYPE AND GENES

As you'll remember, we said that, generally speaking, chromosomes are arranged in homologous pairs. Each member of the pair is *similar* to, but also *different* from, the other member. Human cells, we said, have a total of 23 pairs of chromosomes. We also said that spermatozoa and ova are exceptions: *their* chromosomes *don't* come in pairs and each cell has only 23 chromosomes *total*.

Let's look at one pair of chromosomes and, in particular, let's focus on one part of each chromosome. We want to make the picture easy to look at, so we won't bother with actually shaping the chromosomes like a double helix/twisted ladder.

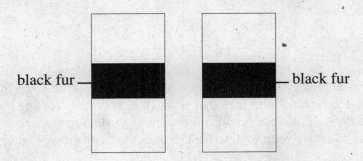

Notice that each of these shaded portions is labeled "black fur." This means that the particular part of the chromosome that we've decided to look at codes for fur color. It's responsible for producing the enzymes that catalyze the chemical reactions that determine the color of the organism's fur. Now, this *particular* pair of chromosomes codes for enzymes that give rise to black fur. So, when it comes to fur color, what is this organism's phenotype? Simple—it's black.

Now, while we're looking at these two chromosomes, let's introduce another word: genotype. As you know, when we say *pheno*type, we're talking about the organism and its traits. But when we say *geno*type, we aren't doing that. Instead, we're talking about the *genes* responsible for those traits. For the organism that we're now talking about—the one with the black fur—we'd (1) look at the homologous chromosome pair pictured above, (2) see that the organism has black fur, and we'd say:

Fur Color *Geno*type: black/black (or BB)

Fur Color *Pheno*type: black.

MORE ABOUT GENOTYPE AND PHENOTYPE: FEATURES THAT ARE DOMINANT AND RECESSIVE

Here's another organism with black fur.

When it comes to fur, what's the organism's phenotype? It's black.

But now let's look at its genotype. Here are the chromosomes that contain the genes responsible for fur color.

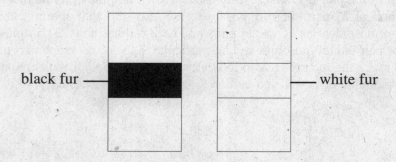

Notice that the two homologous chromosomes don't agree on what color the organism's fur should be. One wants it to be black and the other wants it to be white. Yet one member of the pair definitely wins the debate. That's essentially the meaning of dominance and recessiveness. The organism has *black* fur. When we want to describe this organism's phenotype and genotype in terms of fur color we say:

*Geno*type: black and white (or Bb)
*Pheno*type: black

Why does one of the chromosomes get to express itself in the organism's phenotype while the other one has to keep quiet? Here's the answer: For the species to which this organism belongs, black fur is dominant, and white fur is recessive. That just means that if an organism of this species has one black-fur chromosome and a white-fur chromosome, the organism will have fur that's *black*. Black fur is dominant—a chromosome that codes for black fur dominates a chromosome that codes for white fur. It wins the dispute. It gets to decide what the *pheno*type will be. White fur is recessive—it has to recede—it doesn't get to show itself if its homologous mate thinks the fur should be black. That's what we mean when talk about features that are dominant and recessive.

Let's consider a few more organisms of this species. Here's one with fur color genotype: white/white.

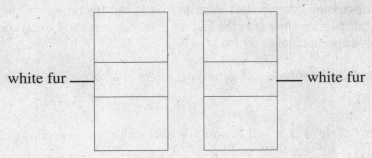

Its fur-color phenotype will be white. Even though white is recessive, both chromosomes agree that the fur should be white. There's no black-fur chromosome around to dominate the matter.

As a matter of fact, if you see an organism of this species and its fur is white, you know that its genotype is bb. Since white is recessive, that's the only genotype that can produce a white phenotype. If a black-fur chromosome were around, the phenotype would have to be black. A white-fur phenotype definitely means a bb genotype since white is recessive.

Here's an organism with fur-color genotype: white/black (bB).

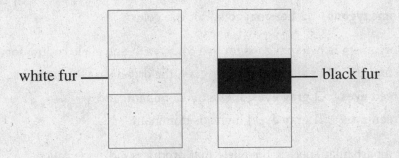

Its fur-color phenotype will most definitely be black. Why? This is because black fur is dominant and white fur is recessive. The black-fur chromosome dominates the white-fur one.

TIME OUT! LET'S LEARN SOME SIMPLE BIOLOGY SUBJECT TEST TERMS: ALLELE, HOMOZYGOUS, HETEROZYGOUS

1. *Allele*:
When biologists refer to a gene that gives rise to more than one version of the same trait—like eye color—they say allele. Just now, for instance, we've been talking about the allele responsible for fur color. The black-fur and the white-fur genes are different alleles for the fur-color gene.

2. *Homo*zygous and *hetero*zygous:
When, for a particular trait, an organism's two alleles are in agreement, we say the organism is *homo*zygous for that trait. The organism that has genotype black/black is homozygous for fur color. The organism that has genotype white/white is also homozygous for fur color. In both cases the two alleles agree on fur color. On the other hand, organisms that had genotypes black/white and white/black were cases where there was a disagreement within the genotype. One allele says white and the other says black. We say that these organisms are *hetero*zygous for fur color, which means that their

two alleles don't agree on what color the fur should be. (We know, of course, that in both cases the fur will be black because, for this organism, black fur is dominant and white fur is recessive.)

We're not done discussing genetics and inheritance yet, but it's time for a quiz. Go back to the beginning of the last discussion (just after the last set of blank lines and boxes) and review. Then fill in the blanks and boxes appropriately.

- The term [❏ **phenotype** ❏ **genotype**] refers to an organism's observable traits.

- The term "allele" [❏ **is** ❏ **is not**] precisely synonymous with the term "gene."

- The term _____ refers, generally, to the fact that the gene responsible for a particular observable trait might exist in more than one version.

- If, in a particular organism, one allele on one member of a homologous chromosome pair codes for blue eye color and a corresponding allele on the other codes for brown eye color, the organism is said to be [❏ **homozygous** ❏ **heterozygous**] for eye color.

- If an organism is heterozygous for eye color, with one allele coding for green and the other allele coding for gray, the organism will have [❏ **green eyes** ❏ **gray eyes**] if *green* is dominant, and [❏ **green eyes** ❏ **gray eyes**] if *gray* is dominant.

- If, for a particular species, the allele that produces a disease called erythemia is dominant and the corresponding allele that produces the absence of disease (a normal organism) is recessive, then:
  (a) an organism with a genotype that is heterozygous for the trait will have the phenotype: [❏ **normal** ❏ **erythemia**].
  (b) an organism with a genotype that is homozygous for the dominant allele will have the phenotype: [❏ **normal** ❏ **erythemia**].
  (c) an organism with a genotype that is homozygous for the recessive allele will have the phenotype: [❏ **normal** ❏ **erythemia**].
  (d) an organism with a phenotype that is normal must have a genotype that is [❏ **homozygous** ❏ **heterozygous**].

MATING AND CROSSING: PREDICTING THE OFFSPRINGS' PHENOTYPES AND GENOTYPES

When you take the Biology Subject Test you'll definitely be asked, in one way or another, to figure out what kind of genotypes and phenotypes to expect from a cross between two organisms that bear offspring. Fortunately, that's easy to do. Suppose two organisms decide to have offspring. One has black fur and the other has white fur. Here are the genotypes:

Male Parent
Bb

Female Parent
bb

Go back to chapter 3 for a few minutes and reread the stuff about meiosis and fertilization (pages 69–74). Realize that, from each pair of chromosomes normally present within the parent cell, the offspring gets one member from its father and another member from its mother. That's because the spermatozoa and ova are <u>hap</u>loid cells, each of which contributes one member of a homologous pair to the zygote which, therefore, is <u>di</u>ploid.

Realize also that any offspring gets its particular mix of chromosomes at random, and different offspring get different mixes. Here's what we mean by that. As for the chromosome that carries the fur-color allele, there are four possible mixtures of homologous chromosomes that a particular offspring might receive from its parents.

1. One animal, for instance, might get this combination from his mother and this one from his father:

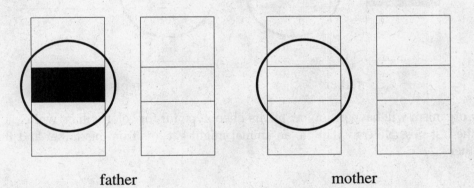

father   mother

That means the new organism will have genotype: Bb. Its phenotype will be black, since black is dominant.

CRACKING GENETICS ◆ 81

2. Another possibility is that the animal might get this from his father and this from his mother:

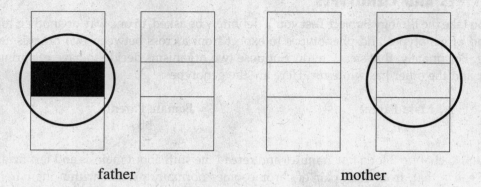

This new organism will also have genotype: Bb. Its phenotype for fur color will be black.

3. In the third possible combination, an animal might get this from his father and this from his mother:

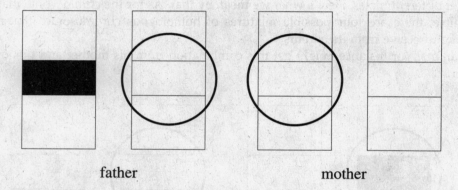

This new organism will have genotype: bb. Its phenotype for fur color will be white.

4. In the last possible combination, an animal might get this from his father and this from his mother:

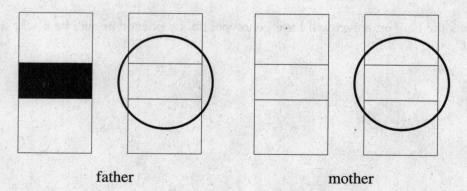

That means the new organism will have genotype: bb. Its phenotype for fur color will be white.

TAKE A LOOK AT THE GENOTYPES AND PHENOTYPES OF THE FOUR OFFSPRING THESE PARENTS MIGHT PRODUCE

If you look back at what we've just done, you'll see that for these parents

- two of the four possible offspring are
  genotype: bb
  phenotype: white

    and

- two of the four possible offspring are
  genotype: Bb
  phenotype: black.

We can say, therefore, that as for *these* parents and the fur color of *their* offspring, the *likelihood* is that

- 50% of the offspring will be:
  genotype: bb
  phenotype: white

    and

- 50% of the offspring will be:
  genotype: Bb
  phenotype: black.

PUNNETT SQUARES

Before we go on with some more examples, we'll show you a little trick. Draw a four-chambered box, and along the sides of the box indicate the genotypes of the two parents, like this:

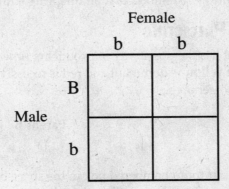

Now, fill in the four segments of the box according to what you've drawn along the sides and you'll easily visualize the four genotype possibilities. (Knowing which trait is dominant and which is recessive means you can easily figure out the phenotype possibilities.) This is called a Punnett square.

```
                    Female
                  b       b
              ┌───────┬───────┐
          B   │  Bb   │  Bb   │
              ├───────┼───────┤
  Male
              │       │       │
          b   │  bb   │  bb   │
              └───────┴───────┘
```

From this, it's quite easy to calculate the likelihood of each phenotype. Once again, you come up with

- 50% offspring of
  genotype: bb
  phenotype: white

  and

- 50% offspring of
  genotype: Bb
  phenotype: black.

What can we say about the likelihood that its fur will be white? Easy. We can say the likelihood is 50%. What can we say about the likelihood that an offspring's fur will be black? Also easy: 50%.

LET'S DO SOME MORE PREDICTING

Suppose some species of plant has the possibility of producing flowers that are (1) red or (2) yellow. Suppose we tell you, also, that yellow is dominant and red is recessive. Now let's contemplate a cross between plants with these two genotypes:

Parent A	Parent B
Yy	Yy

Take a minute and draw your four-chambered box in the space below. Fill it in as we did before. First we'll ask you some questions, and then we'll explain the answers so we're sure you're on top of this stuff.

Refer to your drawing and fill in the blanks and boxes appropriately.

- Parent plant A has phenotype [❑ **red flowers**   ❑ **yellow flowers**].

- Parent plant B has phenotype [❑ **red flowers**   ❑ **yellow flowers**].

- Parent plants A and B [❑ **can**   ❑ **cannot**] possibly produce a plant with red flowers [❑ **because**   ❑ **even though**] both A and B have [❑ **red flowers**   ❑ **yellow flowers**].

- The likelihood that parent plants A and B will produce a plant with red flowers is ____%.

- The likelihood that parent plants A and B will produce a plant with yellow flowers is ____%.

If you draw your box correctly, it probably looked something like this.

	Parent B	
	Y	y
Parent A Y	YY	Yy
y	Yy	yy

You might have drawn it with parent A on top and parent B along the side, or vice versa. It doesn't matter. For each parent you might have arranged the y and Y in either order. That doesn't matter either. You'd still get the same overall results.

To begin with, you know that each parent has phenotype yellow. Yellow is dominant. Even though each parent has a chromosome that wants the flowers to be red, each also has one that wants them to be yellow. Yellow wins the fight in both cases. Both parents have yellow flowers.

Your box shows you that if you make every possible crisscross, your four results are:

1. a plant with
   genotype: YY
   phenotype: yellow

2. a plant with
   genotype: Yy
   phenotype: yellow

3. a plant with
   genotype: Yy
   phenotype: yellow

4. a plant with
   genotype: yy
   phenotype: red

Of the four possible crisscrosses, two give us genotypes that are heterozygous, which means the phenotypes are yellow, one gives us a genotype that's homozygous for the yellow trait which means that it's phenotype is yellow, and one gives us a genotype that is homozygous for the red trait, which means that its phenotype is red.

LET'S DO ANOTHER ONE

Sticking with the red and yellow flowers, let's do some box drawing and make some predictions regarding the offspring of these two parent plants:

Parent A
Yy

Parent B
YY

Draw a box as before. Make the appropriate markings and labels, and then fill in the blank lines and boxes appropriately.

- Parent plant A has phenotype [❏ **red flowers** ❏ **yellow flowers**].

- Parent plant B has phenotype [❏ **red flowers** ❏ **yellow flowers**].

- Parent plants A and B [❏ **can** ❏ **cannot**] possibly produce a plant with red flowers.

- The likelihood that parent plants A and B will produce a plant with red flowers is ____%.

- The likelihood that parent plants A and B will produce a plant with yellow flowers is ____%.

How did you do? Here's the box:

	Parent B	
	Y	Y
Y	YY	YY
y	Yy	Yy

Parent A

One parent is homozygous for the dominant trait (yellow), and the other is heterozygous. For both parents, phenotype is yellow. Your box shows you that the parents might possibly produce these four progeny:

1. a plant with
   genotype: Yy
   phenotype: yellow

2. a plant with
   genotype: Yy
   phenotype: yellow

3. a plant with
   genotype: YY
   phenotype: yellow

4. a plant with
   genotype: YY
   phenotype: yellow

The parents have a 50% likelihood of producing offspring with genotype Yy (or yY, however you want to look at it) and a 50% possibility of producing offspring with genotype YY. In terms of phenotype, they have a 0% chance of producing offspring with red flowers and a 100% chance of producing offspring with yellow flowers. All of their offspring will have yellow flowers. Half will be heterozygous for the trait.

ONE MORE EXAMPLE

Suppose a certain species of animal has the capacity to be born with oily or nonoily feathers. Suppose that oily is dominant and nonoily is recessive. Fill in the blank lines and boxes appropriately.

- If an organism is born with nonoily feathers, it [❏ is   ❏ is not] possible that *both* of its parents were homozygous for oily feathers.

- If an organism is born with oily feathers, it [❏ is   ❏ is not] possible that *both* of its parents were homozygous for oily feathers.

- If an organism is born with oily feathers, it [❏ is   ❏ is not] possible that *one* of its parents was homozygous for nonoily feathers.

- If an organism is born with nonoily feathers and its parent A had oily feathers, then
  (a) the genotype for parent A was _____,
  (b) the genotype for parent B was _____, or _____,
  (c) the phenotype for parent A was _____,
  (d) the phenotype for parent B was _____, or _____.

Got it? Let's see. Here are the answers and explanations:

- If an organism is born with nonoily feathers, it [❏ is   ☑ is not] possible that *both* of its parents were homozygous for oily feathers.

If both parents are homozygous for oily feathers your box would show you that all offspring would also be homozygous for oily feathers. That genotype can't lead to nonoily feathers.

- If an organism is born with oily feathers, it [☑ is   ❏ is not] possible that *both* of its parents were homozygous for oily feathers.

That's because two parents that are homozygous for oily feathers will, in fact, produce offspring that are all oily-feathered. Your box will show you that.

- If an organism is born with oily feathers, it [☑ is   ❏ is not] possible that *one* of its parents was homozygous for nonoily feathers.

That's because an oily-feathered offspring can result from a parent that is homozygous for nonoily feathers so long as the other parent is either heterozygous or homozygous for oily-feathers. Draw a box and you'll see.

- If an organism is born with nonoily feathers and its parent A had oily feathers, then

    (a) the genotype for parent A was <u>Oo</u> (or oO, if you want to look at it that way)

That's because a nonoily-feathered organism *can't* arise if either parent is homozygous for oily feathers. Since you're told that parent A's phenotype is oily, it's got to be heterozygous: Oo.

   (b) the genotype for parent B was <u>Oo or oo</u>

That's because a nonoily-feathered creature can result only if both parents have at least one chromosome for nonoily feathers. That means parent B has to be either Oo or oo.

   (c) the phenotype for parent A was <u>oily</u>

That's because you're told that parent A was oily-feathered.

   (d) the phenotype for parent B was <u>oily</u> or <u>nonoily</u>

That's because a nonoily-feathered creature can result only if *both* parents have at least one chromosome for nonoily feathers. The right answer to item (b) tells you the right answer to this item (d). Parent B has genotype Oo or oo, which means its phenotype is either oily or nonoily.

REMEMBER THIS NAME: GREGOR MENDEL

Almost everything we've just said about dominant traits, recessive traits, and the way they affect offspring was discovered by a guy named Gregor Mendel. When biologists hear "Mendel," they think, father of genetics. In the nineteenth century, Gregor Mendel performed a bunch of experiments on garden peas and that's how he figured out almost everything we've just learned. Sometimes the stuff we've just learned is called Mendelian genetics. Don't forget about Gregor Mendel; he's the father of modern genetics.

Fill in the blanks and boxes appropriately.

- The laws of genetics were drawn up based on the experiments of _____.

- Modern understanding of genetics is attributable largely to the work of
  [❑ Albert Einstein   ❑ Gregor Mendel   ❑ Charles Darwin].

ANOTHER THING ABOUT GENETICS AND INHERITANCE: SEX AND SEX-LINKED TRAITS

Let's think about people. As you know, their cells (other than spermatozoa and ova) have 23 pairs of chromosomes. One of those 23 pairs is called the pair of sex chromosomes. That pair determines whether you're male or female and has loads of genes that determine sexual attributes.

A sex chromosome can either be male, in which case we call it Y, or female, in which case we call it X. A person whose phenotype is female has the sex genotype XX. A person whose phenotype is male has the sex genotype XY. The male got an X chromosome from his mother (whose genotype is XX) and a Y chromosome from his father (whose genotype is XY).

Female children, on the other hand, get an X chromosome from their mothers and an X chromosome from their fathers. So whether a child is born male or female depends on whether it gets from its father an X or a Y chromosome—since the father can donate either one.

Fill in the blanks and boxes appropriately.

- A male person receives from his father a(n) [❏ X   ❏ Y] chromosome.

- A male person [❏ may   ❏ must] have the genotype [❏ XY, ❏ XX or ❏ YY].

- A female person [❏ may   ❏ must] have the genotype [❏ XX, ❏ XX or ❏ XY].

- In terms of sex, all persons male or female receive from their mothers a(n) _____ chromosome.

- In terms of sex, all females receive from their fathers a(n) _____ chromosome.

The sex chromosomes X and Y, as we said, carry certain genes on them. When we say "sex-linked trait," we're talking about a trait whose allele is carried on the X chromosome. So, when you think sex-linked, think <u>carried on the X chromosome</u>. The fact that some traits are sex-linked (carried on the X chromosome) leads to a few interesting results and might easily show up on the Biology Subject Test questions.

Let's take the disease hemophilia. It's a sex-linked trait. In other words, *if* a person carries the trait, he or she carries it on the X chromosome. It's also recessive. We can say hemophilia is a sex-linked recessive trait. What does that mean? It means that a female can only get hemophilia if she's homozygous for the X chromosome that carries the trait for hemophilia. For instance, take a look at these two female genotypes.

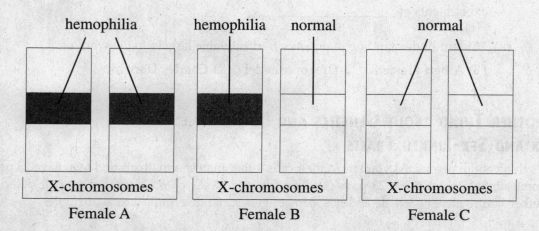

Like all females, Females A, B, and C each have two X chromosomes. For Female A, both X chromosomes carry the hemophilia trait. That means she's homozygous for the disease and her phenotype will be—hemophilia. Female B carries the trait on one of her X chromosomes. That means she's carrying the trait but because it's recessive, she won't show it in her phenotype. She won't have hemophilia. Her phenotype is normal. Female C is homozygous for the normal trait. She's not even carrying the trait for hemophilia. Naturally her phenotype will be normal.

One of the things you just learned—whether you know it or not—is: When it comes to sex-linked, *recessive* traits, a *female* can only express the trait in her phenotype if her genotype is homozygous for it. If she's heterozygous for the trait, then the trait will not be expressed (it won't show up) in her phenotype.

Now, what about males? Let's continue discussing hemophilia, a sex-linked recessive trait, and look at these two male genotypes:

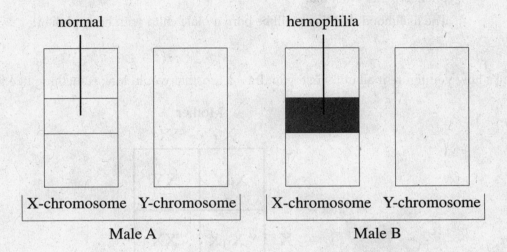

Like all males, Males A and B have an X chromosome (originally received from the mother) and a Y chromosome (originally received from the father). Male B is carrying the hemophilia trait on his X chromosome. Does he have the disease? That is, is his phenotype hemophilia? The answer is yes. Why? Because even though the trait is recessive there's no normal X chromosome around to dominate it. The male doesn't have a second X chromosome. He has a Y chromosome instead and *that* won't suppress the expression of the X chromosome's allele.

So think about it. When it comes to a sex-linked recessive trait, a female will actually express the trait in her phenotype only if she's homozygous for it. A male will express the trait in his phenotype if—plain and simple—he's got it on his X chromosome.

Now, still talking about hemophilia, consider these two parents:

Father
$X_{(normal)}/Y$

Mother
$X_{(hemophilia)}/X_{(normal)}$

Draw a box, and figure out, in terms of sex and hemophilia, what kind of children these two parents might produce. After you've done that, fill in the blanks and boxes appropriately.

- With reference to the parents whose genotypes are shown above,
    (a) ____% of children are likely to be male.
    (b) ____% of children are likely to be female.
    (c) There is [❏ no ❏ some] likelihood that a child will have phenotype: hemophilia.
    (d) There is [❏ no ❏ some] likelihood that a female child will have phenotype: hemophilia.
    (e) There is [❏ no ❏ some] likelihood that a male child will have phenotype: hemophilia.
    (f) The likelihood that there will be born a male child with hemophilia is ____%.

Here's how you figure it all out. First, you draw a box that would look something like this:

	Mother X_H	Mother X
Father Y	$X_H Y$	XY
Father X	$X_H X$	XX

The box shows you that the parents are likely to produce:

1. a male child with
   genotype: $X_H Y$
   phenotype: hemophilia

2. a male child with
   genotype: XY
   phenotype: normal

3. a female child with
   genotype: $X_H X$
   phenotype: normal

4. a female child with:
   genotype: XX
   phenotype: normal

In our talk about sex-linked traits, we've only discussed a sex-linked trait that's recessive. (You probably don't have to worry about a sex-linked trait that's dominant because the Biology Subject Test isn't likely to ask you about one.) Two sex-linked recessive traits that you'll want to keep in mind for the SAT II are hemophilia, which we've already discussed, and color-blindness. Remember these two. The SAT II may not tell you that they are sex-linked; instead, they'll expect you to know that—and now you do.

ONE LAST POINT ABOUT SEX-LINKED RECESSIVE TRAITS: THEY CAN'T GO *FROM* A MALE *TO* A MALE

If the Biology Subject Test presents you with some situation about a sex-linked recessive trait, realize this simple truth: A sex-linked recessive trait can't go from a father to a son. That's because the sex-linked trait is located on the X chromosome and a father does not give his son an X chromosome. He gives him a Y chromosome. Remember that.

5 Cracking the Structures and Functions of Organisms

Nobody understands the structures and functions of *all* organisms. (It's almost impossible to understand the structure and function of *one* organism.)

The Biology Subject Test writers say they'll test you on the structures and functions of organisms, but they really want you to know a *little something* about animal bodies and how they work: bodily support, nutrition, blood, immunity, circulation, respiration, the nervous system, hormones, excretion, reproduction, embryology, behavior, evolution, and ecology.

That sounds like a lot right now, but you'll soon see that it all breaks down to a few important facts per category.

BODILY SUPPORT IN ANIMALS: ENDOSKELETONS AND EXOSKELETONS

Lots of animals have skeletons. The skeleton holds the body together in some recognizable shape. Without a skeleton an organism's body would be, basically, a pile of mush. So, when you see the word skeleton, think shape and support.

Some animals, like human beings, have skeletons made of bones, and they're located inside the body. (Can we see our bones? Are they sitting out in plain view? No. They're sitting *within* the body. When a skeleton sits *inside* the body, it's called an endoskeleton. And when it comes to endoskeletons remember this: *All vertebrates (animals with backbones) have them*. Fish, amphibians, mammals, reptiles, and birds are all vertebrates and they all have endoskeletons.

Fill in the blank lines and boxes appropriately.

- Vertebrates [❑ **do** ❑ **do not**] have skeletons.

- Vertebrates have [❑ **endoskeletons** ❑ **exoskeletons** ❑ **no skeletons**].

- Mammals have [❑ **endoskeletons** ❑ **exoskeletons**].

- Birds have [❑ **endoskeletons** ❑ **exoskeletons**].

Some animals, on the other hand, have skeletons *outside* their bodies. Their skeletons are made not of bones but of hard, crusty shells. Because they're located outside the body, we call them exoskeletons. For the Biology Subject Test you should know that a group of animals called arthropods have exoskeletons. The Biology Subject Test might mention such arthropods as insects, arachnids and crustaceans. Whenever you see mention of arthropods, or any of the animals just listed, or indeed, *any animal with a hard outer shell, think*: exoskeleton. The hard outer shell contains a substance called chitin.

Fill in the blanks and boxes appropriately.

- Vertebrates have an [❑ **endoskeleton** ❑ **exoskeleton**], and arthropods have an [❑ **endoskeleton** ❑ **exoskeleton**].

- Arthropods have an [❑ **endoskeleton** ❑ **exoskeleton**] composed of [❑ **bone** ❑ **chitin**].

- The body of a reptile derives shape and support from an [❑ **endoskeleton** ❑ **exoskeleton**], and the body of an insect derives shape and support from an [❑ **endoskeleton** ❑ **exoskeleton**].

- An endoskeleton is characteristic of _____, and an exoskeleton is characteristic of _____.

Now that we've talked about bodily support we can move on to—nutrition.

WHAT NUTRITION IS ALL ABOUT

Nutrition is about (1) the things an organism needs from the outside world, (2) how it gets them, and (3) what it does with them once it gets it.

Cells need glucose. They need it to conduct glycolysis, the Krebs cycle, and oxidative phosphorylation. Cells also need proteins. They need them to serve as enzymes. Cells also need fat.

How does the body get its glucose? It eats animals or plants and ingests carbohydrate, takes the carbohydrate apart into hydrogen, oxygen, and carbon and then uses the hydrogen, oxygen, and carbon to make glucose.

How does the body get protein? It ingests other animals or plants and their protein, takes them apart into their amino acids, and then uses the amino acids to build the *particular* proteins *it* needs.

How does the body get its fat? Also from animals or plants.

WATER AND VITAMINS

Water is important for nutrition. It isn't food, but the body needs it—lots of it. The body also needs vitamins. They aren't food either. They're found *within* food, and the body definitely needs them: they're important for nutrition.

So—when we think nutrition, we think:

1. Water,
2. Food, which means fats, carbohydrates, protein, and
3. Vitamins, which are contained within food.

For people and lots of other animals, nutrition starts with eating or—to be fancier—feeding. Feeding means sending food, water, and vitamins on a trip through the digestive tract.

A Trip through the Digestive Tract

Here's the human digestive tract:

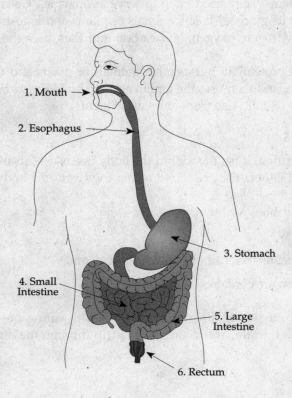

Food goes into the mouth and we swallow it. Then it's in the esophagus. The esophagus wriggles around and pushes the food into the stomach. The wriggling is called peristalsis. Take a look:

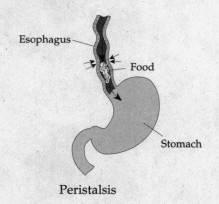

Peristalsis

The esophagus, small intestine, and large intestine all know how to wriggle—they all do peristalsis. Peristalsis sends food that goes from mouth → esophagus → stomach → small intestine → large intestine → rectum.

The large intestine has a nickname: the colon. The digestive tract *as a whole* has a nickname too—the alimentary canal or alimentary tract. When you see alimentary, think "digestive tract."

Reread everything we've said so far about the human digestive tract, then fill in the blanks and boxes appropriately.

- Food travels the human digestive tract by passing through these structures, in this order:
  1. _____,
  2. _____,
  3. _____,
  4. _____ _____,
  5. _____ _____,
  6. _____.

- The word colon is a synonym for [❏ large   ❏ small] intestine.

- The phrase alimentary canal is synonymous with the phrase _____ _____.

What the Digestive Tract Actually *Does*: The Process of *Digestion*

The digestive tract takes food on a journey, and it does things to the food; namely, breaks food down into molecules in the process of digestion.

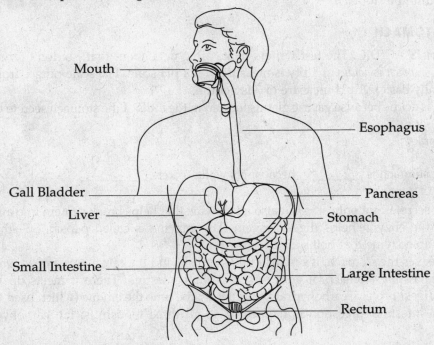

The illustration on the previous page shows only a little bit of the pancreas. That's because the pancreas tucks itself into a loop of small intestine on its left end and then extends to the right behind the stomach where, in this picture, we can't see it. If we were to take the stomach away, we'd see the pancreas tucked into a loop of small intestine like this:

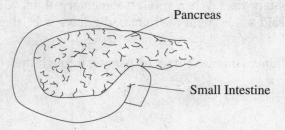

Digestion begins in the mouth, and the mouth has saliva in it. The saliva is released from the salivary glands in the mouth. Another word for release, by the way, is secrete. The salivary glands of the mouth *secrete* saliva.

There's an enzyme in saliva called amylase. Amylase catalyzes a reaction that breaks long carbohydrate molecules into little pieces. Amylase *helps digest starch*. When you see "amylase," think

- digestive enzyme
- contained in saliva, which is secreted in the mouth
- helps digest starch, which is a carbohydrate

The mouth is also important because it has teeth. Teeth chew and turn food into mush that's easy to swallow. Biologists have a fancy word for chewing: mastication.

When food leaves the mouth (1) it's been chewed up into mush which forms a ball, called a bollus and (2) some of its carbohydrate has been partly digested, thanks to amylase. It moves through the esophagus and into the stomach.

INSIDE THE STOMACH

THE STOMACH IS ACIDIC. The test writers make a big deal about that, so don't forget it. Once again, now: *The stomach is acidic.* Acidity is measured on a pH scale. The scale ranges from 1 (really acidic) to 14 (really basic). A pH measure of 7 is neutral.

The stomach is acidic because gastric glands located in the *walls* of the stomach secrete hydrochloric acid (HCl).

Fill in the blanks.

- The stomach is _____, because its walls secrete _____.

The stomach secretes not only acid but also an enzyme that helps break protein into amino acids. In other words the enzyme helps digest protein. The enzyme is called pepsin. So—the stomach secretes pepsin, an enzyme that helps digest protein.

When food leaves the stomach, it's *really* a pile of mush, and the mush gets a new name: chyme. When chyme leaves the stomach, it goes into the small intestine. There it meets up with more enzymes. They digest protein, carbohydrate, and fat. By the time the chyme (which used to be food) enters the small intestine, digestion is happening to all three foodstuffs: fat, carbohydrate, and protein.

Also in the small intestine, the chyme is exposed to a substance called bile. For the Biology Subject Test, you don't have to understand what bile is made of—but you must understand what it does. Bile is made by the liver. Once made, it's sent to the gallbladder for storage. When chyme arrives in the small intestine, the gallbladder sends some bile into the small intestine to greet the chyme.

What does the bile do in the small intestine? Before we answer this, we're going to tell you what it does NOT do: It does NOT digest fat. Bile does not digest fat because it is not an enzyme. The test-writers might make a big deal about this. Now that you know what bile does NOT do, what exactly does it do? Bile emulsifies fats. Emulsifies is just a fancy word for "breaks up." Bile breaks up fat into smaller pieces of fat so that real enzymes can get to them more easily. When it comes to bile, remember:

1. It's made in the liver.
2. It's stored in the gallbladder.
3. It's sent from the gallbladder to the small intestine to emulsify fat.

MORE ABOUT THE SMALL INTESTINE

The small intestine is long—very long. As chyme travels through it, its walls act like a sponge. They absorb all the products of digestion: amino acids, small carbohydrates, and the fatty acids that come from fat digestion. From the walls of the small intestine, all this stuff works its way into blood vessels that run all over the body.

To recap, the small intestine:

1. secretes digestive enzymes

    and

2. absorbs the products of digestion into its walls and then into the blood stream.

THE LARGE INTESTINE

Once the chyme gets through the small intestine, it travels through the large intestine. The *large* intestine mostly reabsorbs water from the chyme. After the chyme gets through the *large* intestine we stop calling it chyme. It's now called feces, and it's eliminated through the rectum.

Review everything we've said about the *process of nutrition*. Then fill in the blanks and boxes appropriately.

- The salivary glands of the mouth secrete an enzyme called _____,
  which functions in the digestion of _____.

- The stomach secretes [❏ **acid**   ❏ **base**], and it is therefore [❏ **acidic**
  ❏ **basic**].

- The stomach [❏ **does**   ❏ **does not**] secrete a digestive enzyme.

- Pepsin is an _____, secreted by the _____. It serves in the
  digestion of _____.

- From the stomach, chyme enters the _____ _____.

- Within the small intestine, additional enzymes [❏ do ❏ do not] appear.

- Enzymes that appear in the small intestine serve in the digestion of _____, _____, and _____.

- The liver produces _____, which is stored in the _____ _____.

- Bile is sent to the small intestine from the _____ _____, and serves to _____.

- The products of digestion are absorbed into the blood stream through the walls of the _____.

- The large intestine's primary function is to _____ _____.

What about Vitamins?

Here's what you've got to know. The human body needs vitamins to keep working well. Remember, we said earlier that vitamins function as coenzymes. The human body only needs vitamins in very small amounts, but it does need them.

When it comes to specific vitamins and the Biology Subject Test remember that:

1. You need **vitamin A** to make a substance called **retinal**, which is necessary for sight. If you don't get enough vitamin A, you may get **night blindness**.

2. You need **vitamin C** to make something called **collagen**. Collagen is a kind of glue that holds your body together. If you don't get enough vitamin C, you get **scurvy**.

3. **Vitamin D** keeps **bones** in good shape, and also helps make them grow. If *kids* don't get enough vitamin D, their bones don't grow correctly. That's called **rickets**.

4. You need **vitamin K** to help form **blood clots** when you get cuts. Without vitamin K, you might bleed to death when you cut yourself.

And vitamin B?

There's no such thing as vitamin B. There are a whole bunch of different vitamins that get *tagged* with the label "B," and we tell them apart by using the names B_1, B_2, B_3, etc. The B vitamins don't have anything to do with one another except that they all have the label B. Each does something entirely different from the others and the Biology Subject Test probably won't ask you what it *is* they do.

NUTRITION IN PLANTS

For the Biology Subject Test, you only have to worry about *green* plants. The main thing you want to know about their nutrition has to do with the way they get their carbohydrates. They get carbohydrates by making them—through the process called photosynthesis.

WHAT IS PHOTOSYNTHESIS?

No one can really understand photosynthesis without first knowing an awful lot about physics and chemistry. Here's what we'll do. We'll tell you everything you have to know about photosynthesis in order to do well on the Biology Subject Test.

Animals get their carbohydrates by eating them. (Then, as we've already discussed, they use their carbohydrates to make ATP from which they draw energy to do all of the things they do.) Plants are different. They don't eat carbohydrates, they *make* them. Specifically, they make glucose, and they make it from water and carbon dioxide (CO_2). **When we say photosynthesis, we're talking about the process by which plants make glucose.**

Glucose has lots of energy. The energy is stored in the chemical bonds that hold the glucose molecule together. Water and carbon dioxide molecules *don't* have lots of energy in *their* chemical bonds. So, if a plant is going to make glucose from water and carbon dioxide, it has to get some energy. It gets that energy <u>from the sun</u> and that's the whole trick to photosynthesis. *Green plants can take energy directly from the sun*. With that energy they can make glucose, a high-energy molecule, from water and carbon dioxide, which are low-energy molecules.

Plants have something in their cells called chlorophyll. Within the plant cell, chlorophyll is located in things called chloroplasts and *those* are among a group of structures called *plastids*. Plant cells and *only* plant cells have chloroplasts. Animal cells don't. Chlorophyll knows how to grab hold of sunlight and take energy from it. Chlorophyll is green (usually). Plants have so much of it that it makes *them* green.

HOW PHOTOSYNTHESIS HAPPENS

There are two reactions in photosynthesis: the light reaction and the dark reaction. Here's what happens in each reaction.

> **Light reaction:** A water molecule is split by sunlight into hydrogen, oxygen, and electrons. Chlorophyll is activated and the electrons are passed down a chain to make ATP and NADPH.
> $ADP + P_i + ENERGY\ FROM\ SUN \rightarrow ATP$
> $NADP + H_2 + ENERGY\ FROM\ SUN \rightarrow NADPH_2$
>
> **Dark reaction:** Glucose is formed using the ATP and $NADPH_2$ from the light reaction, along with CO_2.
> $NADPH_2 + ATP + CO_2 \rightarrow C_6H_{12}O_6$

Through a process known as the Calvin cycle, which occurs in the stroma the cell produces glucose.

The whole photosynthetic reaction (the light and dark phases) can be written like this:

$$6\ CO_2 + 12\ H_2O + \textbf{ENERGY} \rightarrow \textbf{C}_6\textbf{H}_{12}\textbf{O}_6 + 6\ O_2 + 6\ H_2O$$
$$[\text{glucose}]$$

Overall, the cell uses carbon dioxide, water and energy from the sun; it *produces* glucose, oxygen, and water.

WHERE PHOTOSYNTHESIS HAPPENS

Photosynthesis happens in a plant's leaves. There's a lot of room on a leaf's surface for sunlight to hit. Take a look at this cross-section of a leaf:

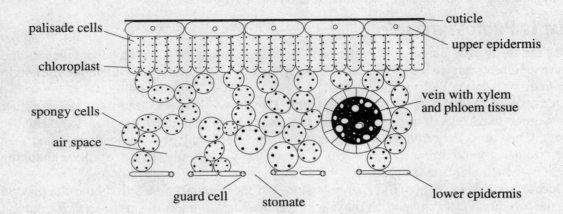

The palisade layer, just under the surface, is where most of photosynthesis takes place. The spongy layer cells beneath the palisade layer also carry out photosynthesis, but this layer is more important for gas exchange. That's why there are so many air pockets in the spongy layer. Now let's consider the top and bottom of the leaf. The outer cell layer of the leaf is called the epidermis (like skin), and the epidermis has a layer of wax on it called the cuticle. The job of the cuticle is (a) to protect the leaf from attack by things like fungi and (b) to keep water from escaping from the leaf. Look at the bottom of the leaf pictured and you'll see something labeled stomate. That's the opening where the leaf sends out and takes in gases like O_2, CO_2, and H_2O. Stomate are opened and closed by—what else—guard cells.

TWO WORDS—AUTOTROPH AND HETEROTROPH

The letters "troph" refer generally to nutrition. Green plants, as we've just explained, can make glucose on their own. That's why we call them autotrophs. ("Auto" means "self.") When we say autotroph, we're talking about an organism that can make its own carbohydrates—which usually means it's an organism that can conduct photosynthesis.

Animals, as you know, can't make their own carbohydrates. They have to get their carbohydrates by eating them. So, we say animals are heterotrophs, which just means they can't do photosynthesis. They have to get carbohydrates from the outside world.

Review everything we've said about photosynthesis. Then fill in the blanks and boxes appropriately.

- Photosynthesis is the process by which green plants *produce* [❑ **water and carbon dioxide** ❑ **glucose**] *from* [❑ **water and carbon dioxide** ❑ **glucose**].

- Photosynthesis consists of (1) a _____ reaction in which energy is taken from sunlight in order to form $NADPH_2$ and ATP, and (2) a _____ reaction in which the energy is used to form _____ from _____ _____ and water.

- The pigment that enables green plants to trap energy from sunlight is called _____, which within the cell is located in the _____.

- When the plant cell conducts the photosynthetic dark reaction, it derives energy from the substances _____ and _____ in order to form glucose from carbon dioxide and water.

- Because green plants can conduct photosynthesis and animals cannot, plants are called [❏ **autotrophs**   ❏ **heterotrophs**], and animals are called [❏ **autotrophs**   ❏ **heterotrophs**].

- Chloroplasts belong to a class of organelles called _____.

- Stomate size is regulated by [❏ **palisade cells**   ❏ **spongy cells**   ❏ **guard cells**].

- Photosynthesis is carried out primarily in the _____ layer of a leaf.

- Gas exchange is carried out primarily in the _____ layer of a leaf.

So much for the topic of nutrition. Now, let's talk about blood and immunity.

BLOOD AND IMMUNITY

Some organisms have blood. Blood carries things all around the body, making deliveries and pickups here, there, and everywhere. For instance, in people, blood makes pickups from the small intestine. The small intestine absorbs digested food and sends it into the blood stream.

Blood consists of two things: (1) fluid and (2) cells that float around in the fluid. The fluid is called plasma and the cells are called blood cells.

BLOOD CELLS: RED AND WHITE

For the Biology Subject Test you'll want to remember that floating around in the plasma are red cells and white cells. Red cells are red, and there are <u>loads</u> of them. There are so many red blood cells that they make blood red.

Red cells contain a protein called hemoglobin. Hemoglobin carries oxygen. Since red cells contain hemoglobin and hemoglobin carries oxygen, we say that red cells carry oxygen and deliver it to cells all over the body. Really, though, it's the hemoglobin that carries the oxygen.

Know this about hemoglobin: It's built partly of iron, which means that if you don't get enough iron, you can't make enough hemoglobin. Your red blood cells then have trouble delivering oxygen. When that happens, we say you have anemia.

For the Biology Subject Test, anemia means:

- not enough iron in the diet
- not enough hemoglobin in the red blood cells and
- not enough oxygen delivered around the body

So much for red blood cells.

White Blood Cells

When we talk about white blood cells, we're really talking about the body's immune system. The immune system helps the body fight germs, which really means viruses, bacteria, and other microscopic organisms.

White blood cells fight viruses, bacteria, and other microorganisms (germs) that cause disease. So remember, white blood cells fight infections.

The Biology Subject Test probably won't talk about other kinds of white blood cells. However, you should know that a white blood cell called a T-lymphocyte is important in fighting infection.

Here's another thing to know: AIDS is a disease that's caused by a virus. The virus somehow takes power away from white blood cells so they can't fight infection. Patients with AIDS die of infections because their bodies can't fight them.

Finally, all blood cells—red and white—are made in the bone marrow. We find bone marrow inside the bones.

Let's review:

- Blood cells are made in the bone marrow.
- Blood cells may be red or white.
- Red blood cells contain hemoglobin, and hemoglobin carries oxygen around the body. (Hemoglobin contains iron.)
- White blood cells serve the immune system; they help the body fight infection.
- The T-lymphocyte is also a white blood cell. It helps fight infection.
- AIDS is caused by a virus that strips white blood cells of their power. AIDS patients have very poor immunity and die of infections.

Now, fill in the blanks and boxes appropriately.

- [❑ Red blood cells   ❑ White blood cells] function in the immune system.

- [❑ Red blood cells   ❑ White blood cells] contain hemoglobin.

- Hemoglobin delivers _____ throughout the body.

- Hemoglobin contains _____ .

- White blood cells serve to fight infection by _____ , and other microorganisms.

- T-lymphocytes are [❑ **white**   ❑ **red**] blood cells.

- AIDS patients have inefficient [❑ **Red blood cells**   ❑ **White blood cells**] and therefore have a defective _____ system.

GAS EXCHANGE AND CIRCULATION

In human beings and other big-time organisms, gas exchange means

- The organism inhales; it takes oxygen into its lungs.
- The walls of the lungs, like sponges, take up the oxygen and pass it into little blood vessels that lie close to the lungs' walls. Those little blood vessels are called pulmonary capillaries.
- Meanwhile, the capillaries deliver something *to* the lungs: water and carbon dioxide.
- Then the organism exhales. It pushes out the water and carbon dioxide that the pulmonary capillaries delivered to the lungs.

Where do the water and carbon dioxide come from? Earlier you learned about the electron transport chain. It *produces* water and carbon dioxide. The cells dump that water and carbon dioxide into the blood. That's how it gets in the blood. The blood carries it to the lungs and that's how it gets to the lungs.

What happens to the oxygen that goes from the lungs into the blood? The blood carries it around, delivering it to every cell in the body. The cells use it in the Krebs cycle and as electron acceptor in the electron transport chain.

LET'S PICTURE ALL OF THIS

Here are the major structures of the respiratory system.

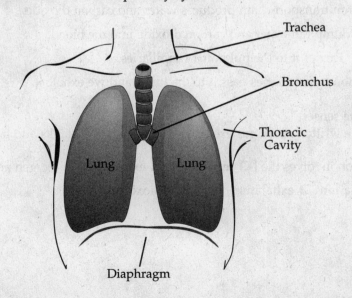

When air is breathed in through the nose or mouth, it passes the trachea, the bronchi, and the bronchioles, and enters the small, dead-end pockets of space.

Look under the microscope at a tiny portion of the lung.

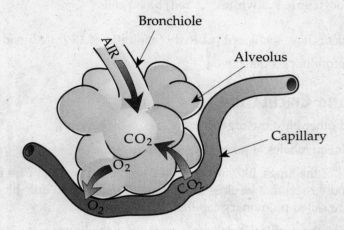

The walls are full of these little pockets. Each one is called an alveolus and alongside each one is a capillary. Oxygen flows into the alveoli when we inhale. Then it crosses from the alveolus into the pulmonary capillary. Carbon dioxide crosses the *other* way—from the pulmonary capillary into the alveolus.

THINGS ARE STARTING TO COME TOGETHER!

Let's recap:

1. We take oxygen into our lungs and our lungs pass it to the blood.
2. The blood sends the oxygen to all of our cells.
3. The cells use oxygen in the Krebs cycle and the electron transport chain to make ATP (which is what they use as their source of energy).
4. The electron transport chain produces water and carbon dioxide.
5. The cells dump the water and carbon dioxide into the blood.
6. The blood carries it to the pulmonary capillaries.
7. The pulmonary capillaries pass it to the lungs and we exhale it.

See? It makes some sense.

Reread what we've written about gas exchange. Then fill in the blanks and boxes appropriately.

- Respiration involves the [❏ inhalation ❏ exhalation] of oxygen and the [❏ inhalation ❏ exhalation] of carbon dioxide.

- Within the lung, oxygen is inhaled and then crosses the wall of the _____ to enter the pulmonary _____.

- Within the lung, _____ _____ crosses the wall of a pulmonary capillary to enter the _____ and then be exhaled.

- The carbon dioxide and water that reach the lung via the blood stream are produced in the body's cells as a result of cellular _____.

WHAT KEEPS THE BLOOD MOVING? THE HEART

The heart is the body's pump. It takes blood in on one side and squishes it out on the other. It keeps blood moving all day, all night, all life long.

The heart is cut into four sections. Two on the left and two on the right. There's a

- right atrium and right ventricle

    and

- a left atrium and left ventricle.

Let's take a look at a rough picture of the heart:

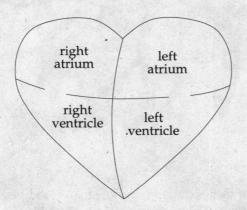

Blood comes into the heart at the right atrium and leaves the heart at the left ventricle. It travels through the body in a big circle. Along the way, it makes deliveries *to* cells and pickups *from* cells.

SOME DETAILS ABOUT THE HEART

Blood leaves the left ventricle and goes immediately into a great big blood vessel called the aorta. The aorta divides over and over again into several little vessels.

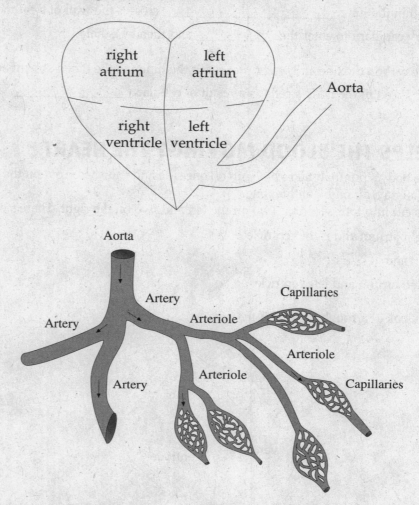

In other words the aorta branches out, into lots of small blood vessels.

The aorta itself is actually a great big artery, but its initial branches are also called arteries. As branches get smaller, they're called arter<u>ioles</u>. When they get *really* small they're called capillaries. Capillaries are extremely small and *their walls are very thin.* Material can actually pass right through capillary walls. Capillaries come in close contact with the body's cells. The capillaries allow the blood and the cells to trade: material *from the blood* passes through the capillary walls and into the cells. Material *from the cells* passes through the capillary walls and into the blood. (You know, for instance, that oxygen moves from blood to cells and that water and carbon dioxide move from cells to blood.) For the Biology Subject Test we say that the capillaries are the site of exchange between blood and cells.

THE CIRCLE CONTINUES

After blood and cells have exchanged materials through the capillary walls, the blood keeps going. Capillaries get *bigger and bigger* and join with other capillaries. A few joined capillaries make a venule, and a few joined venules make a vein. Ultimately, all of the body's veins join together to form two great big veins which are called:

1. the anterior vena cava and
2. the posterior vena cava

The anterior and posterior vena cavae enter the heart's *right atrium*.

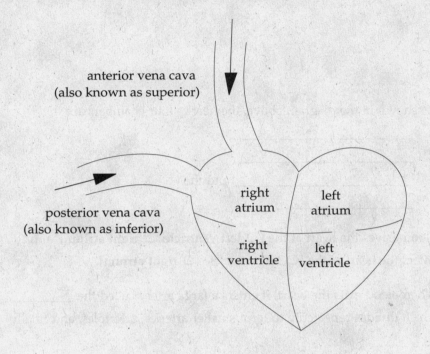

So, the blood:
- goes from the heart's left ventricle all around the body
- makes deliveries and pickups through the capillary walls
- goes back to the heart's right atrium through the superior and inferior vena cavae.

Review everything we've said about circulation. Then fill in the blanks and boxes appropriately.

- As shown in the diagram above, the heart's four chambers are the
  1. _____ _____,
  2. _____ _____,
  3. _____ _____, and the
  4. _____ _____.

- Blood leaves the heart at the [❑ **left ventricle** ❑ **right atrium**], and enters the heart at the [❑ **left ventricle** ❑ **right atrium**].

- When blood exits the heart, it enters a large artery called the _____, which divides repeatedly to form smaller arteries, arterioles, and finally _____.

- The vessels that come into intimate contact with the body's cells in order to conduct exchange of gas and other substances are the [❑ **venules** ❑ **capillaries**].

- Capillaries come together to form _____ and veins, ultimately forming two large veins called the _____ _____ _____ and the _____ _____ _____.

- Blood is conducted back to the heart ultimately by the [❑ **anterior and posterior vena cavae** ❑ **aorta**].

THE PULMONARY CIRCULATION

When blood gets back from its tour around the body, it rides into the right atrium. After the long trip it doesn't have much oxygen left because it's been delivering oxygen (and other material) all over the place. It needs a fresh supply of oxygen, which means it's got to go to the lungs.

The right atrium sends blood straight into the right ventricle and the right ventricle sends it into a thing called the pulmonary artery. The pulmonary artery divides right away into two arteries called the right pulmonary artery and the left pulmonary artery. The right pulmonary artery goes to the right lung and the left pulmonary artery goes to the left lung.

The right and left pulmonary arteries—just like the aorta—divide and divide and divide into tiny little vessels. The tiniest of these vessels are the pulmonary capillaries. We talked before about the pulmonary capillaries when we were discussing gas exchange.

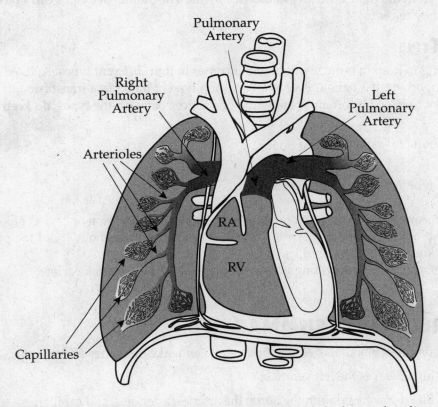

Every little pulmonary capillary gets up close to an alveolus, which we also discussed before. Through the walls of the pulmonary capillaries, the blood

1. *delivers to* the alveolus the carbon dioxide that the blood picked up from the body's cells and

2. *picks up from* the alveolus a fresh supply of oxygen.

Then the pulmonary capillaries get bigger and bigger, forming pulmonary venules. The venules meet and form a large *right* pulmonary vein and a large *left* pulmonary vein. The two pulmonary veins then empty into the left atrium. The blood in the left atrium has dumped its carbon dioxide and water and picked up a fresh supply of oxygen. It's ready for another tour of duty—to make deliveries and pickups all over the body. The left atrium sends the blood directly into the left ventricle and from there the blood goes out into the aorta, as before.

Don't Get Confused about the Pulmonary Artery and Pulmonary Vein

When people think about the pulmonary artery and pulmonary vein, they sometimes forget which one is which. That's because most arteries in the body carry oxygen-*rich* blood. Most veins in the body carry oxygen-*poor* blood. The pulmonary arteries and veins are exceptions. The pulmonary artery carries oxygen-*poor* blood and the pulmonary vein carries oxygen-*rich* blood.

There's an easy way to avoid confusion. Just remember this: If a vessel carries blood away from the heart, it's an artery. If a vessel carries blood into the heart, it's a vein. The pulmonary arteries carry blood away from the right ventricle, so they're *arteries*. The pulmonary veins carry blood into the left atrium, so they're *veins*.

Typical Types

One last thing to keep in mind about blood: it comes in four different types: A, B, AB, and O. Blood types are important. If a patient is given the wrong type of blood in a transfusion, it could be fatal! While it's not terribly important to know the differences between the types, do keep in mind that

- type **O** is the universal donor

- type **AB** is the universal receiver

All this means for us is that anyone can receive a blood transfusion of type O blood, while those with type AB blood (which is very rare among Americans—only about 4% of the population) can receive any kind of blood without risks.

Now that you've got everything you need to know, let's do a quick review.

THE CIRCULATION SUMMARIZED

The circulation takes blood all over the body so it can make deliveries and pickups.

1. Blood leaves the left ventricle.//
2. Blood travels through the aorta, the arteries, arterioles, and capillaries.
3. When blood is in the capillaries—all over the body—it gives things *to* the cells and takes things *from* the cells.
4. Blood passes from capillaries to venules, from venules to veins, and from veins to the anterior and posterior vena cavae (which are just very big veins).
5. Blood goes from the anterior and posterior vena cavae into the heart's right atrium, and then to the right ventricle.
6. The heart's right ventricle sends the blood through the right and left pulmonary arteries into the right and left lungs.

7. The pulmonary arteries divide many times to form zillions of little pulmonary capillaries.
8. Each pulmonary capillary gets close to one of the lung's alveoli.
9. Blood in the pulmonary capillaries delivers carbon dioxide and water to the lungs. The lungs, in turn, deliver oxygen to the blood.
10. The pulmonary capillaries get together to form pulmonary venules, which get together to form pulmonary veins which then get together to form the large right and left pulmonary veins.
11. The pulmonary veins empty into the left atrium, which sends the blood into the left ventricle, which then sends the blood to the aorta, and the whole thing starts over again.

Review everything we've said about pulmonary circulation. Then fill in the blanks and boxes appropriately.

- Blood that enters the right atrium after touring the entire body is [❏ **oxygen-rich**   ❏ **oxygen-poor**].

- From the right atrium, blood is passed immediately to the _____ _____.

- From the right ventricle [❏ **oxygen-rich**   ❏ **oxygen-poor**] blood is passed to the [❏ **pulmonary arteries**   ❏ **pulmonary veins**], and then to the lungs.

- From the lungs, [❏ **oxygen-rich**   ❏ **oxygen-poor**] blood is passed to the [❏ **pulmonary arteries**   ❏ **pulmonary veins**], and then to the heart's _____ _____.

- The pulmonary arteries carry [❏ **oxygen-rich**   ❏ **oxygen-poor**] blood [❏ **away from**   ❏ **toward**] the heart and the pulmonary veins carry [❏ **oxygen-rich**   ❏ **oxygen-poor**] blood [❏ **away from**   ❏ **toward**] the heart.

- The aorta carries [❏ **oxygen-rich**   ❏ **oxygen-poor**] blood [❏ **away from**   ❏ **toward**] the heart and the two vena cavae carry [❏ **oxygen-rich**   ❏ **oxygen-poor**] blood [❏ **away from**   ❏ **toward**] the heart.

Do *Plants* Have a Circulatory System?

Well, yes. Plants don't have blood, but they do have *vessels* through which food and water move. The test writers might want you to know the names of these vessels, so remember them:

1. Vessels in plant tissue called *phloem* carry food like carbohydrates and protein up and down the plant. It's important to know, by the way, that the vessels sitting in the phloem can carry food up *and* down. The test writers might ask about that. Phloem contains two types of cells: sieve tube cells and companion cells. The sieve tube cells are the ones that actually carry out transport. The companion cells help sieve tube cells carry out their metabolic functions.

2. Vessels that sit in a tissue called *xylem* carry *water* up the plant. The water comes from the ground, and the plant's *roots* have the job of getting a hold of it. Xylem tissue is also made up of several types of cells. A fancy name for the cells that directly transport water are tracheids.

(When you're thinking about xylem and phloem, remember that the "ph" sound is like the "f" sound, so that phloem goes with food. Xylem, therefore, must go with water.)

Fill in the blanks and boxes appropriately.

- [❑ Xylem   ❑ Phloem] serves primarily to move water up a plant.

- [❑ Xylem   ❑ Phloem] serves primarily to move carbohydrate and protein up and down a plant.

- Food moves up and down a plant through the _____.

- Water moves through _____ in a plant.

- Vessels located within the [❑ xylem   ❑ phloem] move protein and carbohydrate in the [❑ upward direction only   ❑ downward direction only   ❑ upward and downward directions].

- Vessels that directly conduct organic compounds in a plant are called [❑ tracheids   ❑ sieve tube cells]

Since We Just Mentioned *Roots,* We Might as Well Tell You This

Roots, as we just said, absorb water from the ground (and the water is then moved up and down the plant in vessels of the xylem). Roots, however, absorb more than water. They also absorb nutrients, principally minerals. Finally, remember that roots also help a plant stay put. They give it a home by anchoring it in the ground.

So when you think of roots, think

- absorb water
- absorb nutrients (minerals)
- anchor.

Fill in the blank lines appropriately.

- A plant's roots serve to absorb _____, and _____. They also serve as an _____ that maintains the plant's position.

THE NERVOUS SYSTEM

Every big-time organism (like a human being) has a nervous system. It consists of billions of *nerve cells*. Nerve cells are also called neurons.

A neuron is a cell, but it's put together in a special way. Here's a typical neuron.

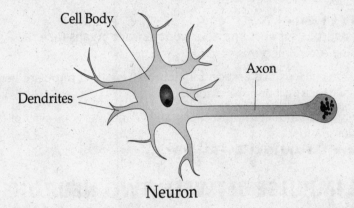

Neuron

The cell body has all the usual cellular material. It's got the nucleus, the Golgi body, the mitochondria, the ribosomes, and all the rest. Neurons are different from other cells because the cell body is connected at one end to dendrites and at the other to an axon.

The Biology Subject Test might show you an *unlabeled* picture like the one above. When you see it,

1. you should know it's a neuron

   and

2. you should be able to label the cell body, dendrites, and axon.

WHAT NEURONS DO

Neurons act like electrical wires, carrying current from one place to another. For the Biology Subject Test we say: Neurons conduct impulses—or—Neurons conduct signals.

Know about the direction in which neurons conduct their impulses. A neuron *receives* an impulse at its dendrites. It transmits the impulse through the cell body and along the axon. The *direction* in which an impulse travels through a neuron is dendrite-cell body-axon. For the Biology Subject Test you must remember that impulses always travel in that direction and ONLY in that direction.

Often waiting at the end of one neuron's axon is another neuron. Its dendrites pick up the impulse and send it along. In other words, an impulse might travel from one neuron's axon to another neuron's dendrites. Then that *second* neuron sends the impulse from its dendrites, to its cell body, and out along its axon.

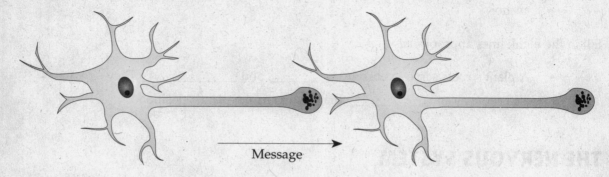
Message

THREE TYPES OF NEURONS YOU'LL NEED TO KNOW ABOUT

The test-writers might ask you to differentiate between three different kinds of neurons. Fortunately, that's easy:

- [*long dendrite, short axon*] <u>afferent (running to)</u> <u>Sensory</u> <u>neurons</u> conduct impulses from sensory organs (like the eyes or skin) to the central nervous system.

- [*short dendrites, longer short axons*] <u>Interneurons</u> (also called association neurons) conduct impulses between sensory neurons and motor neurons.

- [*short dendrites, long axons*] <u>Motor</u> <u>neurons</u> conduct impulses from the central nervous system to muscles or glands. <u>efferent (running away from)</u>

MOVING AN IMPULSE BETWEEN TWO NEURONS: THE SYNAPSE

The axon of one neuron doesn't actually *touch* the dendrites of the next. When an impulse moves from one neuron to the next, it has to cross a little space. That space is called a synapse. For the Biology Subject Test, you should know how the impulse gets across the synapse.

An impulse crosses a synapse by sending a chemical as a messenger. When an impulse reaches the end of one axon, the axon releases a chemical that flows (almost instantly) over to the dendrites of the next neuron. It's the *chemical* that actually gets the impulse started in the second neuron.

What's the messenger's name? For the Biology Subject Test it's acetylcholine. Acetylcholine is a neurotransmitter that (1) gets released from the end of an axon, (2) flows instantly to the dendrites of a neuron next in line, and (3) causes that next neuron to start sending an impulse.

Reread everything we've told you about neurons. Then fill in the blanks and boxes appropriately.

- A nerve cell is also called a _____.

- Within a neuron, the mitochondria, ribosomes, and other organelles are located in the [❏ dendrites ❏ cell body ❏ axon].

- A neuron conducts a nerve impulse in the direction
  [❏ axon →cell body →dendrite ❏ dendrite →cell body →axon].

- Within a neuron, the dendrites carry the nerve impulse [❏ toward ❏ away from] the cell body, and the axon carries the nerve impulse [❏ toward ❏ away from] the cell body.

- If an impulse moves from neuron A to neuron B, then it moves from the [❏ dendrites ❏ axon] of neuron A to the [❏ dendrites ❏ axon] of neuron B.

- A neuron that transmits an impulse from the central nervous system directly to a muscle is called a(n)_____.

- When an impulse moves from one neuron to the next, the neurons [❏ are ❏ are not] in physical contact.

- When an impulse moves from one neuron to the next, it is carried by a chemical _____.

- One of the substances responsible for moving a nerve impulse from one neuron to the next is called _____.

- A neuron that is located between a sensory neuron and a motor neuron is called a(n)_____.

The Nerve Impulse: Polarization, Action Potential, Depolarization, Repolarization

We haven't said anything about *how* a neuron carries an impulse along its dendrites, cell body, and axon from its dendrites to its cell body *itself*. We'll do that now.

But. . .

Don't try to understand this; just remember it!

When a neuron is just waiting around doing nothing, we call it a resting neuron. We also say that a resting neuron is polarized. That means the neuron has lots of *negative* charges *inside* and lots of *positive* charges *outside*.

The positive charge just outside a neuron comes mostly from sodium ions. (Sodium ions are positively charged.)

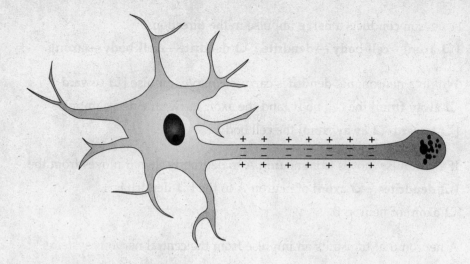

Resting Neuron

Here's what happens when an impulse begins at a neuron's dendrites. All of a sudden, at the tip of the dendrite, a tiny portion of cell membrane becomes very permeable to sodium ions. That makes sodium ions rush into the neuron. So many of them rush in that the cell—*at that little location*—is more positive inside than it is outside. That's quite a change; the whole neuron used to be negative inside and positive outside. We call the change depolarization. Depolarization means one small portion of the neuron is *positive inside* and *negative outside*.

When a cell depolarizes, we think of another phrase: action potential. When we say "action potential," we're talking about the fact that some little section of a dendrite decided to depolarize. (Maybe it got hit with some acetylcholine.) An action potential *causes* the neuron to depolarize,

- which means it becomes very permeable to sodium,
- which means that sodium ions rush inward,
- which means that the little section of neuron becomes positive inside and negative outside.

So remember: *An action potential causes a neuron to depolarize.*
Fill in the blanks and boxes appropriately.

- A resting neuron is relatively positive _____ side and negative _____ side.

- When a neuron experiences an action potential its permeability to _____ increases, which results in _____ization.

- A section of a neuron that is depolarized, due to an influx of _____ ions, is relatively _____ive on the outside and _____ive on the inside.

AFTER DEPOLARIZATION COMES REPOLARIZATION

Once a little bit of neuron depolarizes, it quickly returns to its normal condition: negative inside and positive outside. In other words, it repolarizes. How? Neurons contain some potassium ions. Potassium ions are positively charged. As soon as a little portion of a neuron becomes depolarized because a whole bunch of *sodium* ions rush inward, a whole bunch of potassium ions decide to rush *outward*. When we think about that, we say there's an efflux of potassium ions. The section of neuron that was depolarized is repolarized: it's negative inside and positive outside, just like it was before the action potential.

MEANWHILE, THE ACTION POTENTIAL SPREADS

When one little section of a neuron becomes depolarized and then repolarized, the next little section does the same thing, and then the *next* little section does the same thing, and so on. The action potential spreads along the whole neuron—from dendrites to cell body to axon.

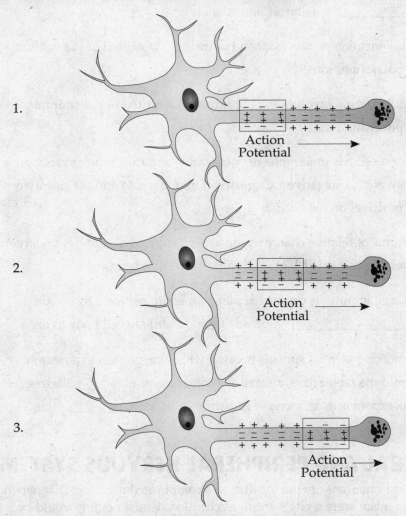

Within fractions of a second, every little area of the neuron—one little section at a time—becomes depolarized and then repolarized.

The spreading action potential *is* the nerve impulse. When you think of a nerve impulse moving from one end of a neuron to another, remember first that a resting neuron is polarized: it's negative inside and positive outside. Then think action potential:

- A small segment of neuron at the dendrite end becomes depolarized and then repolarized because sodium ions rush in and then potassium ions rush out.

- The action potential spreads right along the whole neuron from dendrites to cell body to axon.

Review what we've said about polarization, depolarization and the action potential. Fill in the blanks and boxes appropriately.

- When a portion of a neuron undergoes depolarization, it then undergoes _____ation almost immediately.

- Depolarization results from an [❑ influx ❑ efflux] of [❑ sodium ❑ potassium] ions.

- Repolarization involves the [❑ influx ❑ efflux] of [❑ sodium ❑ potassium] ions.

- Once a cell has undergone depolarization and then repolarization, it is relatively [❑ negative ❑ positive] on the inside and [❑ negative ❑ positive] on the outside.

- In terms of relative charge inside and outside, a cell that has undergone repolarization is [❑ like ❑ unlike] a resting neuron.

- A nerve impulse is conducted along an entire neuron because the _____ _____ spreads from dendrite to cell body to axon.

- An action potential spreads because when one portion of a neuron's cell membrane repolarizes, a small portion adjacent to it ____polarizes and thus experiences an increase in permeability to _____ ions.

THE CENTRAL AND PERIPHERAL NERVOUS SYSTEMS

All of an organism's neurons are put together in a complicated network. That network is the nervous system. If an organism were a city's entire electrical system, a neuron would be a single wire. The nervous system would be all of the wires in the whole city—on every street, on every power line and pole, above ground, below ground, in every wall, of every floor of every building and home. In the human nervous system, billions (**yes, billions**) of neurons run every which way, with synapses all over the place, carrying impulses here, there, and everywhere.

The human being really has only one nervous system. Unfortunately, though, when biologists first studied the human nervous system, they decided to invent the terms central and peripheral. Central nervous system refers to neurons in the brain and the spinal cord. The brain and the spinal cord are absolutely *loaded* with neurons. When we think of all of those neurons together, we say central nervous system. When we think of all of the neurons *outside* the brain and spinal cord—in our skin, our blood vessels, and inside our organs—we say "peripheral nervous system."

The truth is that the central nervous system and peripheral nervous system are hooked up—they're *one* system. Still, when we think of the brain and spinal cord, we're in the habit of saying central. When we think about neurons located outside the brain and spinal cord we say peripheral.

To get it straight, think of a railroad. At the *main* station, there are loads of tracks and loads of trains—arriving, departing, or just sitting around. The tracks lead out in every direction to the towns and villages. Within any town or village, there isn't a whole lot of railroad. There's a track here and a track there. Every now and then a train shows up.

Now imagine we decide to say "central railroad system" when we're talking about the main station and "peripheral railroad system" when we're talking about all of the tracks, trains, and stations in the towns and villages. There would still be only one railroad, but we'd be dividing it in our heads into the main station and the outlying stations.

That's what we do when we think of the nervous system.

- Central nervous system = neurons in brain and spinal cord.
- Peripheral nervous system = all other neurons.

Just as the central nervous system includes the spinal cord and the brain, the peripheral nervous system includes the somatic nervous system and the autonomic nervous system. What's more, the autonomic nervous system includes the sympathetic nervous system and the parasympathetic nervous system. How can you keep all this straight? Look over the following chart and you'll have it down in no time:

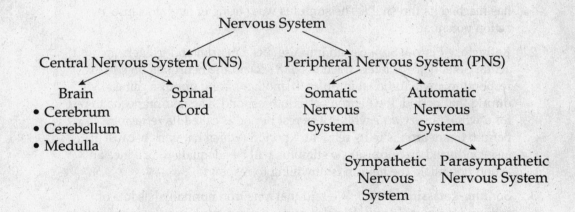

Notice that we listed some structures under the "brain" segment of the CNS. The cerebrum, cerebellum and medulla are all parts of the brain, and each does a different job. The cerebrum takes care of voluntary things, like movement, thoughts, and memory. The cerebellum deals with muscle movement and balance. The medulla handles involuntary acts, like breathing and heart rate.

Review what we've said about the central and peripheral nervous systems. Then fill in the blanks and boxes appropriately.

- The phrase "_____ nervous system" refers to neurons located *within* the brain and spinal cord.

- The medulla is responsible for coordinating [❏ **thought processes** ❏ **breathing rate**].

- The phrase "_____ nervous system" refers to neurons located *outside* of the brain and spinal cord.

- The cerebellum helps an organism to maintain its [❏ **heart rate** ❏ **balance**]

- Neurons of the peripheral nervous system are [❏ **entirely separate from** ❏ **connected to**] neurons of the central nervous system.

- Thought processes are carried out by the [❏ **medulla** ❏ **cerebellum** ❏ **cerebrum**].

Some More Biology Subject Test Words: Threshold, Refractory Period, Sodium-Potassium Pump, Myelin Sheath, Schwann Cells

Here are some more words and phrases to know when you think about the nerve cells:

1. **Threshold:** Suppose a neuron's dendrites get hit by some impulse. If the impulse is strong enough to cause an action potential, we say the neuron has reached its threshold. The stimulus was strong enough to cause an action potential.

2. **Refractory Period:** Suppose an impulse hits a neuron and triggers an action potential. For a few fractions of a second the neuron is unable to respond to any additional impulse. If another strong impulse came along during that period, the neuron wouldn't respond. The neuron needs to rest for a while (a very *little* while). The rest period is called its refractory period. If a neuron is in its refractory period (which happens because it just conducted an impulse), *no* stimulus will be adequate to produce an action potential. The neuron is unwilling to respond.

3. **Sodium-Potassium Pump:** We said that a neuron normally has lots of sodium ions outside of it. We also said a neuron has a supply of potassium inside. During an action potential, sodium rushes in (depolarization) and potassium then rushes out (repolarization). Suppose that kept happening over and over again. The neuron would eventually have lots of sodium inside and lots of potassium outside. It wouldn't be able to undergo action potentials anymore. The neuron has a simple solution to that problem.

When it's at rest, the neuron pumps sodium out and pumps potassium in. When we think of that process, we say "sodium-potassium" pump.

The sodium-potassium pump requires energy. The test-writers think that's a big deal, so remember it. The neuron's sodium-potassium pump is energy dependent. (The energy, of course, comes from ATP, which, in turn, comes from glycolysis, the Krebs cycle, and oxidative phosphorylation.)

4. **Myelin Sheath and Schwann Cells:** Some (not all) neurons are *wrapped* in a myelin sheath. A neuron with a myelin sheath can conduct impulses extra rapidly. (*All* neurons conduct impulses very quickly. But those with myelin sheaths conduct them even more quickly.) A myelin sheath is actually made of a whole bunch of cells called Schwann cells. Schwann cells wrap themselves around (some) neurons and they *are* the myelin sheath. So, a neuron *is* a cell (as we already know) and its axon can then be wrapped around by another cell, the Schwann cell. The Schwann cell wrapper is the myelin sheath.

Review what we've said about threshold, refractory period, sodium-potassium pump, and myelin sheath. Then fill in the blanks and boxes appropriately.

- If a stimulus reaches a neuron, it [❑ **will**   ❑ **will not**] necessarily produce an action potential.

- Only a stimulus that reaches the neuron's _____ will cause the neuron to experience an action potential.

- If a neuron experiences a stimulus that fails to reach its threshold, it [❑ **will**   ❑ **will not**] experience an action potential.

- For some short period of time after a neuron experiences an action potential, it undergoes a _____ period which means it cannot, for that period, experience an additional action potential no matter how strong an impulse it may receive.

- During a neuron's refractory period the neuron [❑ **can**   ❑ **cannot**] experience an action potential.

- The sodium-potassium pump helps the neuron maintain a high concentration of _____ium internally and a high concentration of _____ium externally.

- Operation of the sodium-potassium pump [❑ **does**   ❑ **does not**] require energy.

- [❑ **All**   ❑ **Some**] neurons are wrapped in a myelin sheath.

- A myelin sheath [❏ **speeds** ❏ **slows**] the neuron's conduction of an impulse.

- A myelin sheath is actually composed of _____ cells.

HORMONES AND THE ENDOCRINE SYSTEM

In complex organisms (like human beings), some organs make chemicals and then secrete (release) them into the blood stream. The chemicals are hormones. The organs that make them are endocrine glands. *Endocrine glands make hormones and secrete them into the blood stream.*

Once a hormone is in the blood stream, it goes everywhere. But each hormone has only some cells that pay attention to it. In other words each hormone has its *effect* only on some cells. When we think of some particular hormone and the cells it affects, we say "target." Suppose gland X makes hormone Y and hormone Y has some effect on organ Z. We say that for hormone Y, Z is the target organ.

What Hormones Do

Each hormone does its own job on its target in its own way. For the Biology Subject Test, we must learn about a couple of different endocrine glands, the hormones they secrete, and the jobs these hormones perform on their targets.

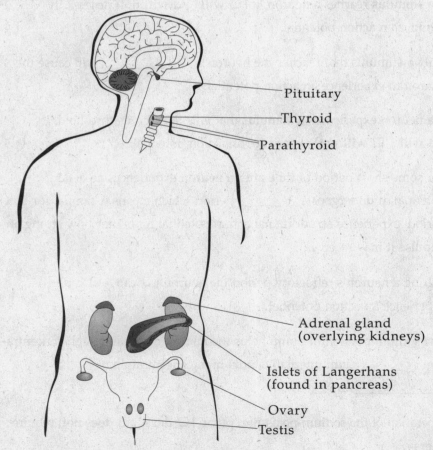

Male and Female Organs are shown

THE PANCREAS

1. **The Islets of Langerhans and their Hormones:** The islets of Langerhans are groups of cells that live in the pancreas, which is located in the abdomen. The islets of Langerhans secrete two hormones; **insulin** and **glucagon**. Let's tackle insulin first. Insulin has an effect on almost every cell in the body. *It makes the cells let glucose in.* If there's no insulin running around in the body, the cells won't take in glucose. Glucose will float around in the blood but it won't get into the cells. The cells need glucose in order to produce ATP (from glycolysis, the Krebs cycle, electron transport, and oxidative phosphorylation). Insulin opens the cells' glucose gates.

The fact that insulin lets glucose move from the blood into the cells also means it reduces the concentration of glucose in the blood. When you think of insulin, think: a hormone that lowers glucose levels in the blood. Remember, though, that the reason insulin lowers blood glucose levels is that it allows glucose into the cells.

If a person's pancreas doesn't secrete enough insulin, that person has diabetes (which is actually called diabetes mellitus). Patients with diabetes have lots of glucose running around in their blood but very little gets into their cells. This is because their pancreas isn't secreting insulin.

Glucagon (the other hormone the islets of Langerhans secrete) has the opposite effect. Glucagon causes the liver to release glucose into the blood stream. Why the liver? Because the liver stores glucose—in the form of glycogen—until it is needed.

So what happens to the blood sugar level when glucagon is around? It goes up. When insulin is around? It goes down.

THE ADRENAL GLAND: MEDULLA AND CORTEX

The adrenal glands are located on top of each kidney. Even though each adrenal gland sits on top of a kidney, the adrenal glands are not part of the kidney. They have two parts: the adrenal medulla (the inner part) and the adrenal cortex (the outer part).

2. **The Adrenal Medulla and its Hormones:** The adrenal medulla secretes **epinephrine** and **norepinephrine**. Just to make things more complicated, these two substances have alternative names. Epinephrine's trade name is adrenaline, and norepinephrine's trade name is noradrenaline. So, the adrenal medulla secretes two hormones:
    (a) Epinephrine (also called Adrenaline) and,
    (b) Norepinephrine (also called Noradrenaline).

These two hormones have similar effects. They also have lots of targets. Both affect the heart by raising its rate. Both affect blood vessels by causing them to narrow (constrict). And both increase the rate of respiration.

Epinephrine's and norepinephrine's effect is to get lots of blood (and hence lots of oxygen) to the body's muscles. When you're about to take on a tough physical activity, the adrenal medulla secretes epinephrine and norepinephrine. That gets you ready for action.

When people think about the adrenal medulla's hormones, they like to say "fight or flight." The expression "fight or flight" might show up on the Biology Subject Test. It's supposed to mean that epinephrine and norepinephrine get you ready for heavy duty physical action by raising your heart rate, constricting your blood vessels, and raising your breathing rate.

We've talked about the adrenal *medulla*. Now it's time to talk about the adrenal *cortex*.

3. **The Adrenal Cortex and its Hormones:** The adrenal cortex secretes a whole bunch of steroid hormones called **corticosteroids**. Corticosteroids have lots of targets and lots of effects. Here's what you need to know.

Some corticosteroids are called **glucocorticoids**. Their target is the liver. The glucocorticoids tell the liver to produce and release glucose into the blood. That means more glucose will run around the blood stream. Assuming there's enough insulin around, it will get into the rest of the body's cells. When you think glucocorticoids, think:

- secreted by the adrenal cortex
- liver is the target
- causes liver to produce and release glucose.

Now we've already told you that the liver stores glucose in the form of glycogen, and that glucagon causes the liver to release the stored glucose. So do glucocorticoids tell the liver to do the same thing that glucagon tells it to do? The answer is no. Glucagon says to the liver, "give up some of the glucose that you're storing as glycogen, so that it's available in the blood stream." Glucocorticoids say to the liver, "make up a batch of glucose from some fat and protein components, and then send it out into the blood stream." So glycogen and glucocorticoids both act on the liver, causing it to release glucose into the blood stream. But they give the liver different commands about where the glucose should come from.

The adrenal cortex also secretes some hormones called **mineralocorticoids**. Their target is the kidney. Mineralocorticoids make the kidney excrete less volume as urine than it would otherwise excrete. Another way of saying this is to say that "mineralocorticoids cause the kidney to retain water." When you think mineralocorticoids, think:

- secreted by the adrenal cortex
- kidney is the target
- causes kidney to retain water.

Just how do mineralocorticoids cause the kidney to retain water? By causing them to retain salt: Na (sodium) and Cl (chloride). The minute the kidneys retain salt, they also automatically retain water. So the mineralocorticoids really cause the kidneys to retain salt, which makes them retain water, too.

4. **The Thyroid Gland and its Hormone—Thyroid Hormone:** The thyroid gland is located in the neck. It secretes **thyroxine**. Thyroxine affects most of the body's cells. It makes them increase their rate of metabolism. That means they work harder and use more energy. If you think of the body as a car, thyroxine steps on the gas.

Thyroxine has iodine in it. If you don't eat enough iodine you can't make enough thyroxine. If that happens you develop hypothyroidism. ("Hypo" means not enough.) If someone is iodine deficient, she develops hypothyroidism and her metabolic rate is low; she can gain weight and become sluggish. (Hyperthyroidism, on the other hand, is due to *over*-production of thyroxine and produces a higher metabolic state, accompanied by symptoms such as weight loss and weakness.)

5. **The Parathyroid and its Hormone—Parathormone:** There are four parathyroid glands. They're really small and they're located near the thyroid gland. The parathyroid glands secrete a hormone called **parathormone**. It's also called **parathyroid hormone**. Parathormone makes bone cells release calcium. The calcium ends up in the blood.

It's important for the body to keep a pretty constant concentration of calcium in its blood so that it's available to cells. The parathyroid gland helps the body do that. If calcium concentration gets too low, it secretes parathormone and the bone releases calcium. When you think of the parathyroid gland and its hormone, think: releases calcium from bone. The calcium is used for lots of things, ranging from nervous impulse conduction to growth of teeth and bones.

The Pituitary Gland: Anterior and Posterior

In the brain, there's a thing called the pituitary gland. It's actually two separate glands that happen to lie right next to each other. The two glands are called the anterior pituitary gland and posterior pituitary gland. They have absolutely nothing to do with each other, but they live so close together that they *look* like one structure.

6. **The Anterior Pituitary Gland and its Hormones:** The anterior pituitary secretes more than one hormone. Right now we'll mention three.
   (a) One of the anterior pituitary's hormones is called **growth hormone**. Its target is all tissues and it causes the body to *grow*. If a child doesn't secrete enough growth hormone, he doesn't grow. Sometimes growth hormone can be injected into children who aren't growing in order to stimulate growth.
   (b) Another of the anterior pituitary's hormones is known as adrenocorticotropic hormone, abbreviated **ACTH**. Its target is *another endocrine gland*—the adrenal cortex. ACTH makes the adrenal cortex go into action. When the anterior pituitary secretes ACTH, the adrenal cortex gets turned on. It starts producing and secreting glucocorticoids and mineralocorticoids.
   (c) The anterior pituitary also secretes a hormone called **thyroid-stimulating hormone**. It's like ACTH because its target is another endocrine gland—the thyroid gland. When the anterior pituitary secretes thyroid stimulating hormone, the thyroid gland gets moving and starts producing and secreting thyroid hormone.

The anterior pituitary secretes some other hormones. We'll mention them in a little while when we discuss the female menstrual cycle.

When you think of the anterior pituitary, remember: growth hormone, ACTH, and thyroid-stimulating hormone.

7. **The Posterior Pituitary Gland and its Hormones:** The posterior pituitary secretes two hormones. One is called antidiuretic hormone (ADH), or vasopressin. The other is called oxytocin.

ADH affects the kidney. Like the mineralocorticoids, ADH causes the kidney to concentrate its urine. It causes the kidney to retain water.

Oxytocin affects the female uterus. When a mother is about to give birth, the posterior pituitary secretes oxytocin. That causes the uterus to start contracting, which pushes the baby through the birth canal. Oxytocin also causes the mammary glands to release milk.

When you think posterior pituitary, remember:

- ADH causes the kidney to retain water
- Oxytocin causes the uterus to contract—promotes birth; causes the mammary glands to release milk

An Annoying Fact about the Posterior Pituitary Hormones

ADH and oxytocin aren't *made* in the posterior pituitary. They're made in another part of the brain—the hypothalamus. The hormones are sent *from* the hypothalamus *to* the posterior pituitary. They stay in the posterior pituitary until they're needed. Then the posterior pituitary secretes them. When it comes to ADH and oxytocin, the hypothalamus does the manufacturing. The posterior pituitary does the secreting.

What Else About the Hypothalamus?

The hypothalamus is also in the brain and releases hormones that have the anterior pituitary as their target organ. These hormones inhibit or stimulate the release of anterior pituitary hormones.

Go back and read everything we've said about hormones secreted by the islets of Langerhans, adrenal medulla, adrenal cortex, thyroid, parathyroid, anterior pituitary, and posterior pituitary glands. For each gland, know the names of the hormones we've discussed, their targets, and their effects. Then take this little matching quiz.

- On each blank line place the number that designates the appropriate hormone.

    A. The islets of Langerhans secrete ____ and ____.

    B. The adrenal cortex secretes ____ and ____.

    C. The anterior pituitary secretes ____, ____, and ____.

    D. The thyroid gland secretes ____.

    1. parathyroid hormone
    2. thyroid-stimulating hormone
    3. epinephrine
    4. growth hormone
    5. insulin
    6. thyroxine
    7. mineralocorticoids
    8. norepinephrine
    9. ACTH

E. The parathyroid gland secretes _____.

F. The posterior pituitary gland secretes _____ and _____.

G. The adrenal medulla secretes _____ and _____.

10. glucocorticoids
11. oxytocin
12. vasopressin
13. glucagon

On each blank line, place the number that designates the appropriate hormone.

Hormone

parathyroid hormone _____

thyroid-stimulating hormone _____

growth hormone _____

insulin _____

thyroid hormone _____

mineralocorticoids _____

ACTH _____

glucocorticoids _____

oxytocin _____

vasopressin _____

epinephrine _____

norepinephrine _____

glucagon _____

Effect

1. increases body's metabolic rate
2. increases blood glucose concentration
3. increases blood calcium concentration
4. increases reabsorption of sodium and water in the kidney
5. decreases blood glucose concentration
6. causes uterus to contract and mammary glands to secrete milk
7. stimulates thyroid gland to produce and release thyroxine
8. causes kidney to retain water
9. targets all tissues and causes tissue growth
10. stimulates the adrenal cortex
11. prepares body for "fight-or-flight" response.

THE FEMALE MENSTRUAL CYCLE

The Biology Subject Test might ask about the human female's menstrual cycle.

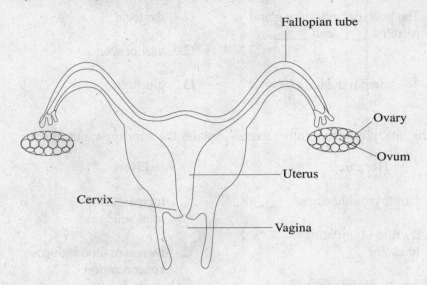

To start, remember that the ovaries are the female's gonads

- They manufacture ova and
- *They're also endocrine organs*, since they secrete estrogen and progesterone—sex hormones that are predominant in females.

Inside the ovary are many oocytes—the cells that eventually become ova. Each oocyte surrounds itself with a bunch of cells. The oocyte plus its surrounding circle of cells is called a *follicle*.

Female mammals and birds are born with all their oocytes. Each oocyte in its follicle is immature, however. At puberty the anterior pituitary begins to release hormones that allow a follicle to mature and produce an ovum.

Now for the menstrual cycle. It's a cycle of hormone secretion. Get to know its key players: (1) the ovaries and (2) the anterior pituitary gland. The menstrual cycle has three phases.

PHASE 1: THE FOLLICULAR PHASE

In phase 1, the anterior pituitary secretes two hormones: follicle-stimulating hormone (FSH) and luteinizing hormone (LH). These stimulate follicles in the ovary to grow. *So during phase 1, follicles in the ovary are growing*. In fact, phase 1 is actually called the follicular phase.

The growing follicle (in the ovary) secretes a hormone too. It secretes estrogen. During the follicular phase, a total of three hormones are secreted:

1. FSH from the anterior pituitary
2. LH from the anterior pituitary

    and

3. estrogen from the ovary (or, more precisely, from the follicle growing *in* the ovary).

Fill in the blanks and boxes appropriately.

- The female menstrual cycle is mediated, chiefly, by secretions from the _____ gland and the _____.

- The anterior pituitary [❏ **does** ❏ **does not**] secrete estrogen.

- The anterior pituitary [❏ **does** ❏ **does not**] secrete luteinizing hormone.

- A follicle consists of an _____ and surrounding layers of _____.

- The ovary/growing follicle [❏ **does** ❏ **does not**] secrete estrogen.

So during the follicular phase, one follicle in the ovary is maturing, and something else also happens: the endometrium (the tissue that lines the uterus) is getting thicker; it's building itself up in preparation for implantation.

The follicular phase lasts about 10 days. Just before it's over, the anterior pituitary suddenly secretes a *very large* amount of LH. When that happens, we call it a luteal surge.

The luteal surge makes the follicle release the ovum, and the ovum enters the fallopian tube (also called the oviduct). In other words, *the luteal surge produces ovulation*—the release of an ovum from the follicle. When the ovum is released, we're at the end of the follicular phase.

Fill in the blanks and boxes appropriately.

- At the end of the follicular phase of the female menstrual cycle, the _____ _____ gland suddenly increases its production of luteinizing hormone.

- The follicular phase of the female menstrual cycle [❏ **begins** ❏ **ends**] with a surge in the secretion of luteinizing hormone (LH).

- The luteal surge causes _____.

- The process by which a mature follicle ruptures and releases its ovum is called _____.

- One follicle matures during a period called the [❏ **luteal surge** ❏ **follicular phase**].

- During the follicular phase, the endometrial lining of the uterus [❏ **builds up** ❏ **sloughs off**].

Phase 2: The Luteal phase

When the follicular phase ends, an ovum is moving through the fallopian tube in the oviduct and a broken follicle is left behind in the ovary. The broken follicle gets a new name; it's called the corpus luteum. The corpus luteum, still sitting in the ovary, becomes an endocrine tissue: it starts to secrete two hormones: (1) estrogen and (2) progesterone.

Progesterone continues getting the uterus ready for pregnancy. Even if the ovum *isn't* fertilized and there *isn't* going to be any pregnancy, the corpus luteum thinks there will be. Progesterone makes the uterus develop a bunch of glands and blood vessels that get it ready for pregnancy.

After about 13 days into the luteal phase (23 days total), if the fertilization and implantation has not occurred, the corpus luteum *stops* secreting estrogen and progesterone.

Fill in the blanks and boxes appropriately.

- During the second phase of the female menstrual cycle, the corpus luteum secretes _____ and _____. The pro_____ gets the uterus ready for pregnancy.

- During the luteal phase of the female menstrual cycle, the ovum, traveling in the fallopian tube, [❑ does   ❑ does not] secrete estrogen and progesterone.

- During the luteal phase of the female menstrual cycle, the corpus luteum, situated in the ovary, [❑ does   ❑ does not] secrete estrogen and progesterone.

Phase 3: Flow

Once the corpus luteum stops secreting estrogen and progesterone, the uterus sheds its new tissue, glands and blood vessels. That causes menstrual bleeding—the female's menstrual period. The bleeding lasts about 5 days. That completes the cycle:

10 days (follicular phase) + 13 days (luteal phase) + 5 days (flow) = 28 days.

Draw lines connecting the event on the left with the appropriate order of occurrence on the right.

Event	Order of Occurrence
Uterus sheds; menstrual bleeding occurs	FIRST
Anterior pituitary secretes FSH and LH; follicle is stimulated and secretes estrogen	SECOND
Corpus luteum secretes estrogen and progesterone	THIRD

- When the corpus luteum stops secreting estrogen and progesterone, the female experiences her menstrual _____, which is really a period of _____ from the uterus.

- The menstrual period (flow phase) of the female menstrual cycle [❏ **is** ❏ **is not**] initiated when the anterior pituitary stops secreting luteinizing hormone.

- The menstrual period (flow phase) of the female menstrual cycle [❏ **is** ❏ **is not**] initiated when the corpus luteum stops secreting estrogen and progesterone.

- When fertilization and implantation occur, the corpus luteum [❏ **does** ❏ **does not**] continue to secrete estrogen and progesterone, and the uterine lining [❏ **is** ❏ **is not**] shed.

A Word about Males

We know that each testis has seminiferous tubules in it. Spermatozoa are formed within the seminiferous tubules. In the testes there are some cells that *aren't* part of the seminiferous tubules. These other cells secrete a hormone called *testosterone*. Testosterone is the male sex hormone. In boys, the testes secrete only a little bit of testosterone. But when a boy gets to be about 13, the testes start to secrete lots of testosterone. Just as FSH and LH cause a follicle to mature and to secrete estrogen in females, FSH and LH cause the testes to produce sperm and secrete testosterone in males. The testosterone gets into the blood and travels all over the body. The testosterone causes the male sex organs and the muscles to develop. When you think testosterone, also think male secondary sex characteristics. That's just a fancy term for saying that testosterone makes men have deep voices and facial hair. When that happens, we say that the boy has reached puberty.

- The principle male sex hormone is called _____ and it's secreted by the _____.

- Testosterone promotes the development of the male sex organs, muscles, and _____ _____ _____, such as facial hair.

Now that we've talked about hormones, let's discuss reproduction and embryology.

HOW HUMAN BEINGS ARE FORMED: REPRODUCTION AND EMBRYOLOGY

Human reproduction and embryology are all about the way human beings form. *Understanding* these subjects is tough. Fortunately you just need to know about a couple of simple terms and events.

What Happens:

(a) **Gametes are formed:** The male produces a sperm cell and the female produces an ovum (egg cell). As we said, the sperm and ovum are both called gametes.

(b) **The ovum is fertilized and we get a zygote:** The ovum travels through the fallopian tube, where it meets up with the sperm. The sperm fertilizes it. Once that happens, we don't have two gametes anymore. Instead we have one structure—a fertilized ovum—which is also called a zygote. The zygote is going to develop, ultimately, into a baby. But first it has to grow.

(c) **Cleavage:** The zygote starts dividing and dividing and dividing—again and again and again. All of this dividing is called cleavage.

(d) **Implantation in uterus:** After the zygote undergoes cleavage, it implants itself in the uterus. There it begins to grow and develop.

(e) **The embryo forms:** From conception (fertilization) until the second month, the developing zygote is called an embryo. It looks something like a cross between a fish and human being.

Once the fetus is fully developed, it's born.
For the Biology Subject Test, therefore, we think of the stages of embryology like this:

Gametes → Fertilization → Zygote → Cleavage → Implantation in Uterus → Embryo → Fetus → Birth.

Don't forget the chick embryo

Sometimes the test writers will ask questions about chick embryos; why chick embryos? Because chick embryos give rise to four extra embryonic membranes in addition to the three germ layers. These membranes are the yolk sac (food for the embryo), the amnion (the membrane with fluid to protect the embryo), the chorion (the membrane that encloses the other membranes, and the allantois (the membrane that stores wastes). Don't forget that.

Fill in the blank lines and boxes.

- The ovum is produced by the [❏ uterus ❏ ovary].

- Implantation in the uterus occurs [❏ before ❏ after] fertilization.

- Fertilization takes place in the [❏ ovary ❏ fallopian tube ❏ uterus].

- The gametes arise [❏ before ❏ after] the zygote arises.

- When the sperm cell fertilizes the ovum, it produces a _____.

- The embryo arises [❑ **before** ❑ **after**] implantation in the uterus.

- Correctly order the following sequence of events and structures associated with human embryology.

    a. fertilization   _____
    b. embryo   _____
    c. gametes   _____
    d. ovulation   _____
    e. implantation in uterus   _____
    f. zygote   _____
    g. birth   _____
    h. cleavage   _____

LET'S GO BACK TO CLEAVAGE FOR A MINUTE

We said that a zygote undergoes cleavage to become an embryo. Actually, there are a bunch of steps in between zygote and embryo.

When you think of a developing embryo, keep this sequence in mind:

1. Zygote
2. Cleavage
3. Blastula
4. Gastrula
5. Embryo

You already know what a zygote is (a fertilized egg) and what cleavage is (a series of cell divisions that the zygote immediately undergoes). Now let's tackle the blastula and gastrula. A blastula is a ball of cells (one cell layer thick) whose center is filled with fluid. A gastrula is that same fluid-filled ball of cells, but with an indentation on one of its sides.

Fill in the following blanks and boxes appropriately:

- A developmental structure that follows cleavage, but precedes the gastrula is the [❑ **zygote** ❑ **embryo** ❑ **blastula**].

- A developmental stage marked by a series of rapid mitotic divisions is known as _____.

- A sphere of cells one layer thick and filled with fluid is called the [❑ **blastula** ❑ **gastrula** ❑ **embryo**].

- A developmental stage known as gastrulation produces a structure that [❏ **is perfectly spherical**   ❏ **bears a depression on one side**].

- The correct sequence of a developing embryo is zygote, _____, _____, _____, and embryo.

During gastrulation (the process of forming a gastrula) the ball of cells forms three layers, called germ layers. These germ layers ultimately give rise to different kinds of tissues in the fully developed person.

The three layers: The names of the three layers all end in the letters "derm." They're <u>ecto</u>derm, <u>meso</u>derm, and <u>endo</u>derm. Here's what you should know about the three germ layers and the structures they ultimately produce in the fully formed person.

1. **The ectoderm:** The ectoderm produces the epidermis (outer layer of skin), the nervous system, and the eye.

2. **The endoderm:** The endoderm produces the inner linings of all of the digestive and respiratory structures. The inner linings of the mouth, nose, trachea, lungs, esophagus, intestine, liver, and gall bladder—all come from endoderm.

3. **The mesoderm:** Everything else comes from mesoderm. That includes bones, muscles, heart, and blood vessels.

It seems complicated, but it's really pretty simple. When you think of the three germ layers just associate:

ectoderm *with*	epidermis, eye, and nervous system
endoderm *with*	inner linings of respiratory and digestive structures
mesoderm *with*	everything else, including bones, muscles, heart, blood vessels and cells, reproductive organs and cartilage.

Fill in the blanks and boxes appropriately.

- The three embryonic germ layers are called _____, _____, and _____.

- The human eye develops from the germ layer known as [❏ **mesoderm**   ❏ **ectoderm**].

- Human blood vessels develop from the germ layer known as [❏ **mesoderm**   ❏ **ectoderm**].

- The *outer* lining of the human intestine develops from the germ layer known as [❑ mesoderm ❑ endoderm].

- The *inner* lining of the human trachea develops from the germ layer known as [❑ ectoderm ❑ endoderm].

- Ectoderm gives rise to the human _____, _____ system and the [❑ outer layer ❑ inner layers] of the skin, known as the _____ .

- Endoderm gives rise to the inner linings of the human _____ and _____ structures.

WHAT ABOUT REPRODUCTION IN PLANTS?

Good question. The test writers might want you to know something about the way *flowering* plants (angiosperms) reproduce. A plant that has flowers, fruits, and seeds is called an angiosperm. So, here's a typical flower:

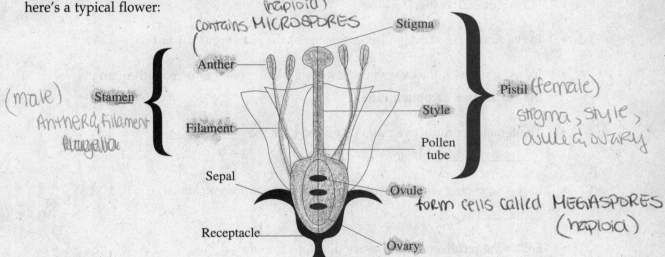

Here's what you should know about the flower's parts and the role they play in reproduction.

The Stamen: The stamen is the plant's male component. It consists of the anther and the filament. The anther makes pollen. Pollen contains little cells called microspores, which you should think of as the flower's version of sperm cells. Like sperm cells, microspores are haploid.

The Pistil: The pistil is the plant's female component. It consists of the stigma, style, ovule, and ovary. Inside the ovary there are ovules, which form little cells called megaspores. You should think of a megaspore as the flower's version of the ovum. Like the ovum, a megaspore is haploid.

Here's how the flowering plant reproduces.

1. Some pollen (containing microspores) somehow falls onto the stigma, which is sticky. Once that happens, the pollen grains germinate at the stigma.

2. As you can see, the stigma is connected to the ovary by the style. During germination, a tube, called a pollen tube, grows down through the style and meets up with the ovary.

3. Microspores (from the pollen) meet up with megaspores (from the ovule). Microspores fertilize the megaspores and they produce not a zygote, but a seed. Once the ovule is fertilized, the ovary develops into fruit.

4. Somehow or other the seed is released. When it finds a suitable environment, it develops into a new plant.

That's all you should remember about the reproduction of plants:

Pollen (microspores) located in anther → Pollen lands and germinates on stigma → Pollen tube grows within the style → Microspores pass down the style and meet up with megaspores (from the ovule) in the ovary → Fertilization → Seed.

Fill in the blanks and boxes appropriately.

- In the flowering plant, pollen is *produced by and located on* the [❏ stigma ❏ anther].

- In the flowering plant, pollen *germinates* at the [❏ stigma ❏ anther].

- In order to reach the ovary in a flowering plant, pollen must pass through the [❏ anther ❏ pollen tube].

- The female reproductive parts of a flower are the _____, _____, and ovary.

- In the flowering plant, microspores are housed within a substance called _____.

- Following fertilization, the ovary develops into the _____.

- In the flowering plant, megaspores are housed within a structure called the _____.

- In the flowering plant, reproduction requires fertilization of _____ by _____, and results in the formation of a _____.

- The male reproductive components of a flower are the _____, _____, and _____.

Now that we've discussed reproduction, let's talk about learning, behavior, and coexistence.

LEARNING, BEHAVIOR, AND COEXISTENCE

The test writers want you to know something about the way animals learn. In particular, you should know about imprinting, conditioning, and insight.

1. INSTINCT:

The first type of behavior we need to keep in mind is inherited behavior, or *instinct*. All organisms, high and low alike, possess some basic instincts. In fact, since instinct underlies all other behavior, it can be thought of as the circuitry that guides behavior.

For example, hive insects such as bees and termites never *learn* their roles; they are born knowing them. On the basis of this inborn knowledge, or instinct, they will carry out all their other behaviors: a worker carries out worker tasks, a drone, drone tasks, and a queen, queen tasks.

2. IMPRINTING:

Think of a gosling (a baby goose). Imagine it's 1 day old and hasn't yet seen its mother. Suppose, then, that its mother walks in front of it and gives out its usual gosling call. Somehow or other the little gosling will decide that this creature is its mother. It will follow this creature and treat it as its mother from then on. That's good, since the creature *is* its mother.

Now think of the same little gosling, but suppose its mother doesn't walk in front of it. Imagine, instead, that *you* walk in front of it and play a *tape* of a gosling's call. Guess what! The gosling will think *you* are its mother.

Believe it or not, the gosling will follow you and will treat you as its mother from then on. It turns out that a one-day-old gosling thinks the first moving thing that walks past it and makes a gosling's call is its mother. When that happens, we call it "imprinting."

Here's the whole point about imprinting. Young animals are impressionable, and during the early days of their lives they use certain cues to make important decisions. For the gosling that identifies its mother, the cue is the fact that something passes in front of it and sounds a gosling call. Somehow or other, that *imprints* on the gosling the idea that the object is its mother.

If the test writers refer to some animal making a decision—*any* decision—early in life on the basis of something it sees, hears, or smells—think imprinting.

Fill in the blank lines and boxes appropriately.

- When an animal makes a decision early in life based on some critical cue,

   it has undergone a learning process called _____.

- If, by repeated trial a cat comes to know that if it rings the bell tied to its

   owner's front door, the owner will open the door, it [❏ has

   ❏ has not] learned by imprinting.

3. CONDITIONING:

Imagine some fish in a tank. If we tap on the left side of the tank, the fish swim away to the right. But suppose as soon as the fish swim away we drop some fish food into the water on the left side of the tank, where we knocked. Suppose, furthermore, that we do this every day for several days: When it's time to feed the fish we knock on the left side of the tank and drop food in on the left side of the tank at the same time.

After several days the fish seem to learn that the knock on the tank means food is coming. They don't swim *away* from the knock, they swim *toward* it. That kind of learning is called conditioning. When we say "conditioning" we're talking about the fact that an animal learns to associate things with each other. Our fish learn to associate the knock with the food. Because conditioning involves associations, it's also called associative learning.

Fill in the blank boxes appropriately.

- If a dog is scolded each time it steps on a living room carpet and for that reason the dog stops stepping on the carpet, then the dog [❏ has ❏ has not] experienced conditioning.

- If a rat is rewarded each time it presses a particular bar and for that reason presses the bar repeatedly, the rat [❏ has ❏ has not] experienced conditioning.

4. INSIGHT:

When we say "insight learning" we're talking about the highest form of learning. Insight means the ability to approach new situations and somehow figure out how to deal with them. As animals go, human beings are pretty good with insight. Another word for "insight learning" is "reasoning."

Fill in the blank boxes appropriately.

- Consider a dog tied to a tree. In its travels it circles the tree clockwise several times and finds that its leash has been shortened because it is wrapped several times around the tree's trunk. The animal has never faced the situation before, but determines to walk counterclockwise around the tree several times in order to restore its leash to its original length. The animal has demonstrated [❏ imprinting ❏ conditioning ❏ reasoning/insight].

- After each description of learning identified below, write:

  (**A**) if it represents imprinting.
  (**B**) if it represents conditioning.
  (**C**) if it represents insight.

1. A dog learns that if it brings the newspaper into the house each evening it gets a bone. _____

2. A man wishes to turn a screw and, having no screwdriver, realizes that he can use the edge of a dull knife as a substitute. _____

3. A bird is born and sees, in its nest, other birds. The bird acts as though, from then on, it belongs to the same species as these other birds. _____

4. A cockroach learns to run from light because every time a light comes on someone tries to step on it. _____

BEHAVIOR THAT *DOESN'T* COME FROM LEARNING: *INSTINCT*

If a horse hears a loud noise, he runs, and learning has nothing to do with it. Some force inside the horse—some voice with which the animal is born—says "run when you hear a loud noise." When an animal demonstrates a behavior that is *not* learned, we call it instinct or innate behavior.

The fact that a baby sucks from a nipple represents instinct. So does the fact that it swallows the milk that comes into its mouth. When an animal is born with an inclination to behave in a certain way, we call that behavior instinct.

BEHAVIOR IN PLANTS: THE TROPISMS

Plants need light. Suppose a plant sits in a room with sunlight coming through a window. The plant will, on its own, bend toward the window. When a plant bends toward light, we call it phototropism.

Furthermore, plant stems like to grow upward. If you take a potted plant and lay it on its side, the stem will bend and the plant will start to grow up—away from the earth. When we think of a plant doing that—growing away from the earth—we call it negative geotropism.

Plants need minerals and water from the earth. Plant *roots*, therefore, like to grow downward, toward the earth. Once again, if we lay a plant on its side its roots will bend downward and grow *toward* the earth. When roots grow toward the earth—it's called positive geotropism.

Within the earth, plant roots tend to grow toward a water supply; this is called hydrotropism. So:

Phototropism *means* growth of the stem towards light.

Negative geotropism *means* growth of the stem upward—away from the Earth.

Positive geotropism *means* growth of the root downward—toward the Earth.

Hydrotropism *means* growth of the root toward a water supply.

All of these tropisms are caused by plant hormones called auxins. For the Biology Subject Test you need not know exactly how auxins work. Just know that (1) auxins are plant hormones and (2) that they induce the growth behaviors we've just discussed:

Phototropism, geotropism (negative and positive), and hydrotropism are all caused by *auxins*.

Fill in the blanks and boxes appropriately.

- If the stem of a plant grows toward a light source it exhibits the phenomenon of [❑ **negative geotropism** ❑ **phototropism**], which is caused by a class of hormones called _____.

- If the root of a plant grows toward a water source it exhibits the phenomenon of [❑ **hydrotropism** ❑ **negative geotropism** ❑ **positive geotropism**], which is caused by a class of hormones called _____.

- If a tree should fall but remain alive and start growing toward the sky, it would be exhibiting the phenomenon of [❑ **hydrotropism** ❑ **negative geotropism** ❑ **positive geotropism**], which is caused by a class of hormones called _____.

- If a tree should fall but remain alive and its roots should insist on growing straight down, it exhibits the phenomenon of [❑ **hydrotropism** ❑ **negative geotropism** ❑ **positive geotropism**], which is caused by a class of hormones called _____.

ANIMALS AND PLANTS CAN TELL TIME: BIOLOGICAL CLOCKS

Think of a plant. Imagine it predictably opens its leaves at 6 AM, closes them at noon, opens them again at 6 PM, and closes them again at midnight. Imagine, furthermore, that the plant does this even if it is kept in the dark all day and all night. This plant seems somehow to know when to open and close its leaves, even without being exposed to changing conditions of sunlight. Something in the plant keeps time.

We don't know how plants, animals or individual cells keep time, but they do seem to do it. In other words, living things have biological clocks. We imagine that somewhere in the cell of the plant or animal there's a clock that tells it what time of day it is, or what time of year it is. The behavior that arises from biological clocks is instinctive. It isn't learned.

Here's a Biology Subject Test term that has something to do with biological clocks: circadian rhythm. When a biological clock makes an organism do something on a *daily* basis, we think "circadian rhythm." The plant that opens and closes its leaves predictably several times daily exhibits a circadian rhythm. The plant that loses its leaves in the fall and regrows them in the spring does *not* reflect a circadian rhythm. The pattern is seasonal, not daily.

Fill in the blank lines and boxes appropriately.

- The fact that plants (or animals) have an innate sense of time is attributable to a "_____ _____."

- When a biological clock dictates behavior conforming to a *24-hour day*, the clock is said to reflect a _____ rhythm.

- Certain mammals predictably hibernate for certain months of the year and do not hibernate in others. Such behavior [❏ **does** ❏ **does not**] reflect a circadian rhythm.

- Certain birds remain in their nests throughout the day and emerge only at night. Such behavior [❏ **does** ❏ **does not**] reflect a circadian rhythm.

ORGANISMS COMMUNICATE: PHEROMONES

The test writers want you to know that most organisms communicate. Know in particular that lots of animals communicate by releasing pheromones. For the Biology Subject Test, a pheromone is any chemical that (a) is released by one member of a species and (b) affects the behavior of other members some useful and predictable way.

Suppose an ant discovers a food source and wants others of his colony to know about it. He lays down a trail of chemicals, leading from the ant colony to the food source. Somehow, the ants know to follow the *chemical* trail and so they are led to the food. The chemical is a pheromone.

Some female animals release chemicals that attract males. Those chemicals are also pheromones. On discovering danger, some animals release chemicals that signal others of their species to stay away. Those chemicals, too, are pheromones. In some species, an animal's dead body releases a chemical that causes survivors to bury it. That chemical is a pheromone as well.

Fill in the blanks and boxes appropriately.

- A chemical (released by an animal) which, in turn, affects the behavior of another animal of its species is a _____ .

- If an animal produces and secretes into its own bloodstream some chemical that produces an effect on *its own* behavior, that chemical [❏ **is** ❏ **is not**] a pheromone.

- If, within a species, one individual produces and secretes a chemical that produces an effect on another, that chemical [❏ **is** ❏ **is not**] a pheromone.

PLANTS AND ANIMALS LIVE TOGETHER: SYMBIOSIS

Organisms of different species sometimes share living spaces; in arrangements called symbiosis. Moss grows on tree trunks. That's symbiosis. Bacteria live in our intestines. That's symbiosis too. For the Biology Subject Test you should know a few words that describe the ways in which symbiotic organisms get along.

Word 1—Mutualism: When two symbiotic organisms confer some benefit on each other, we say they exhibit mutualism.

Word 2—Commensalism: When we say commensalism we mean that:
- one organism benefits from another and
- the other organism doesn't care. It gets no benefit and suffers no harm.

Lots of times, for instance, little plants grow on the branches of big trees. The trees aren't helped or harmed. The little plants, called epiphytes, do get a benefit however: the tree is their home base and elevates them so that they are exposed to sunlight, for instance. That's commensalism.

Word 3—Parasitism: Sometimes one organism lives with another and *does* cause it harm (by feeding off of it, for example). The organism that's doing the harm is called a parasite and we call the relationship—parasitism.

Fill in the blanks and boxes appropriately.

- Whether they harm or help one another, two organisms of different species that are living in close association are said to be in a _____ relationship.

- When two organisms live in intimate association and each confers a benefit on the other, the two organisms are in a _____ relationship that is _____.

- The term symbiosis [❏ **does** ❏ **does not**] describe any arrangement in which two organisms of different species live in close association.

- Mutualism [❏ **is** ❏ **is not**] a form of symbiotic relationship.

- Commensalism [❏ **is** ❏ **is not**] a form of symbiosis.

- If two organisms live in intimate association and one gains a benefit while the other suffers harm, the two organisms [❏ **do** ❏ **do not**] exhibit commensalism.

- If two organisms live in intimate association and one gains a benefit while the other suffers harm, the two organisms [❏ **do** ❏ **do not**] exhibit parasitism.

- Parasitism [❏ **is** ❏ **is not**] a form of symbiotic relationship.

NOW WE'LL TALK ABOUT MICROORGANISMS: PROTISTS, FUNGI, BACTERIA, AND VIRUSES

The Biology Subject Tests want you to know a few things about protists, fungi, bacteria, and viruses. We'll start with protists. These organisms are eukaryotes. Know that:

1. all protists are single-celled and
2. possess a true nucleus with a nuclear membrane.

Remember the names of these three protists:

1. Amoeba, the shapeless blob
2. Paramecium, which has tiny hairs or cilia all over it, and
3. Euglena, which moves about thanks to a flagellum.

Some protists are plant-like and some are animal-like, which means that some can conduct photosynthesis and some can't. But one thing is for sure: *All protists are single-celled.*

Fill in the blanks and boxes appropriately.

- [❏ **Some** ❏ **All**] protists can conduct photosynthesis.

- [❏ **Some** ❏ **All**] protists resemble animal cells.

- [❏ **Some** ❏ **All**] protists are single-celled.

- Fish [❏ **are** ❏ **are not**] protists.

- The euglena [❏ **is** ❏ **is not**] a protist.

- The _____ and the _____ are protists.

- The _____, _____, and _____ are [❏ **multicellular** ❏ **unicellular**].

LET'S TALK ABOUT FUNGI

When you look at a fungus, it seems as though someone built it by taking a lot of cells, putting them together and then taking away the walls *between* the cells. A fungus is like many cells all joined together in one. A fungus has many copies of the usual organelles and it has many nuclei. That's why it's sometimes called a multinucleate cell.

One way fungi reproduce is by forming spores. Do you have to know what a spore is? Nope, but when it comes to fungi and reproduction, think spores.

One group of fungi are called yeasts. Yeasts are exceptions to the spore rule. They reproduce by budding. When the organism sheds a piece of itself and the piece forms a whole new organism, this is budding. When the Biology Subject Test asks about the reproduction of yeasts, just think: budding.

- A fungus has [❏ **one** ❏ **many**] nuclei.

- Most fungi can reproduce by forming spores, but yeasts reproduce by _____.

LET'S TALK ABOUT BACTERIA

A bacterium is a single-celled organism, but it's different from the cells we've talked about up until now. A bacterium is a *prokaryotic* cell, meaning that it has a plasma membrane and a chromosome, but none of the other cellular equipment we're used to. It has no mitochondria, no endoplasmic reticulum, no Golgi bodies, and no nucleus or nuclear membrane. A prokaryotic cell is a primitive kind of cell—like a car without any options, gadgets, or special features. Another thing you need to know about bacteria is that they have a cell wall.

Fill in the blank lines and boxes appropriately.

- Bacteria are _____ karyotic cells.

- Bacterial cells [❏ do   ❏ do not] have mitochondria.

- Bacterial cells [❏ do   ❏ do not] have a Golgi apparatus.

- Bacterial cells [❏ do   ❏ do not] have a cell wall.

Bacteria have chromosomes. The typical bacterium has 1 ring-shaped chromosome, and it's made of DNA. At some point in its life, the bacterium replicates the whole chromosome, which means it has two chromosomes that are exactly alike. Then, a little later the bacterium divides into two daughter bacteria, each of which gets one of the chromosomes. Each daughter ultimately replicates the chromosome, and divides into two new bacteria. When bacteria divide, biologists call the division *fission*. Binary fission means a bacterium divides into 2 equal parts. The cycle continues—one bacterium leading to two, two leading to four, four leading to eight . . . on and on.

Bacteria don't keep reproducing *exact* copies of themselves over and over again. In three different ways they mix their chromosomes with those of other bacteria and so achieve genetic recombination.

Way 1—Transformation: Transformation means that one bacterium gives some of its chromosome to another. The second bacterium ends up with a different genotype and phenotype than it originally had. That second bacterium is transformed.

Way 2—Conjugation: Conjugation means that a bacterium donates *some* of its DNA to another bacterium via a bridge called a pilus.

Way 3—Transduction: Transduction means that a virus takes some DNA from one bacterium and carries it to another.

Reread what we've said about the ways bacteria reproduce. Then—

- By inserting the appropriate number, match the term on the left with the corresponding process on the right.

Conjugation: #_____

1. Virus carries DNA from one bacterium to another

Transduction: #_____

2. One bacterium transfers genetic material to another, causing a change in genotype and phenotype in the second bacterium

Transformation: #_____

3. One bacterium donates a fragment of DNA to another via a pilus.

How Bacteria Get Nutrition

Some bacteria can perform photosynthesis. They're called autotrophs. Most bacteria can<u>not</u> perform photosynthesis. They're heterotrophs.

Some bacteria *absorb* their food from the bodies of other organisms, and some absorb their food from dead bodies (they're saprophytes). Still others absorb their food from living bodies (they're parasites).

Fill in the blanks and boxes appropriately.

- [❑ No bacteria   ❑ Some bacteria] are carnivores.

- [❑ No bacteria   ❑ Some bacteria] are herbivores.

- [❑ No bacteria   ❑ Some bacteria] are autotrophs.

- [❑ Few bacteria   ❑ Most bacteria] are heterotrophs.

- [❑ No bacteria   ❑ Some bacteria] are saprophytes.

- [❑ No bacteria   ❑ Some bacteria] are parasites.

Most bacteria need oxygen. They're <u>obliged</u> to have oxygen, just as we are, and we call them <u>obligate</u> aerobes. Yet, some bacteria can't stand oxygen—it poisons them. We call those bacteria obligate <u>an</u>aerobes.

Finally, there are some bacteria that don't *mind* being in the presence of oxygen, but they don't *need* it. They can take it or leave it. We call these bacteria facultative anaerobes.

On each blank line, insert the appropriate number.

Obligate anaerobe #_____
1. derives energy via aerobic respiration only; cannot survive without oxygen

Facultative anaerobe #_____
2. derives energy via fermentation only; cannot survive in the presence of oxygen

Obligate aerobe #_____
3. can derive energy via aerobic respiration or fermentation; can survive with or without oxygen

Some bacteria do something we call "fixing nitrogen," and here's what you have to know about it. Plants need nitrogen. There's lots of nitrogen in the air, but plants don't know how to get it from the air. They have to get it from the soil. The problem is that some soil is nitrogen-poor. Fortunately for the plants, some bacteria that live in soil do know how to get nitrogen from the air; they are nitrogen-fixing bacteria. Plants that live near nitrogen-fixing bacteria don't have to get their nitrogen from the soil. They get it from the bacteria (who get it from the air).

The plants that have this special association with nitrogen-fixing bacteria are called legumes. Pea plants are legumes, and so are clover plants. The nitrogen-fixing bacteria live in the root nodules of these plants. The association is mutualistic: everyone is happy. The bacteria get shelter, and the legumes get nitrogen in a form they can use to make protein. This association is a favorite of the test writers. When you see "nitrogen-fixing bacteria" think, "legumes . . . root nodules . . . mutualism . . . nitrogen source for plants."

WE STILL HAVE TO TELL YOU ABOUT VIRUSES

A virus isn't really considered a cell. It has only two components: (1) a coat made of protein (only protein) and, inside the coat, (2) nucleic acid. Depending on the virus, the nucleic acid might be DNA or RNA. But get this straight: *all* viruses have some kind of nucleic acid inside their protein coats.

Suppose a virus wants to reproduce. It has none of the equipment a cell normally needs for this work—no centriole, no spindle apparatus—nothing except nucleic acid. It *can't* reproduce without the help of some other cell. So, when a virus wants to reproduce, here's what it does:

First: It latches on to some other cell. It might latch on to a bacterial cell, fungal cell, plant cell or animal cell. We call the other cell the host.

Second: Once the virus has connected to its host, it injects its nucleic acid into the host.

Third: The virus's nucleic acid *uses the host cell's equipment* to reproduce itself—many, many times.

What we end up with then, is many, many new viruses *inside* the host cell. When all is said and done, the virus's nucleic acid has replicated itself many, many times—inside the host cell—and many new virus particles (virions) are formed.

Fourth: The new viral particles rip the host cell open—they "lyse" it—and burst out to freedom. We're left with loads of new virions and (usually) a dead host.

Fill in the blank lines and boxes appropriately.

- A viral coat is made of [❏ **protein only** ❏ **lipid and protein**].

- [❏ **TRUE** ❏ **FALSE**] All viruses contain DNA or RNA.

- [❏ **Some viruses** ❏ **All viruses**] contain nucleic acid.

- Viruses [❏ **do** ❏ **do not**] have, within them, all of the cellular machinery necessary for their own reproduction.

- [❏ **TRUE** ❏ **FALSE**] In order to reproduce themselves, viruses make use of cellular machinery belonging to a host cell.

- In reproducing itself a virus injects its _____ into a host cell.

- Which of the following correctly orders the events associated with viral reproduction?
    - (A) 1. Viral nucleic acid is replicated using host cell machinery
        2. New viral particles lyse host cell and emerge
        3. Protein coat attaches to host cell
        4. Virus injects nucleic acid into host cell
        5. New protein coats are formed
    - (B) 1. Protein coat attaches to host cell
        2. Virus injects nucleic acid into host cell
        3. Viral nucleic acid is replicated using host cell machinery
        4. New protein coats are formed
        5. New viral particles lyse host cell and emerge
    - (C) 1. New protein coats are formed
        2. Virus injects nucleic acid into host cell
        3. Viral nucleic acid is replicated using host cell machinery
        4. New viral particles lyse host cell and emerge
        5. Protein coat attaches to host cell
    - (D) 1. Protein coat attaches to host cell
        2. Virus injects nucleic acid into host cell
        3. New viral particles lyse host cell and emerge
        4. Viral nucleic acid is replicated using host cell machinery
        5. New protein coats are formed

EVOLUTION AND ECOLOGY

Think about a big population of people. Then think about *all* the genes of *all* the people in the *whole* population, this is the population's gene pool.

Within the population, some people have genes for dark skin and others for light skin. Some have an inborn gift for music and some for running quickly. Some have genes for brown eyes and some for green eyes. Each person in the population has a distinct set of genes, different from all others. This is called genetic variability.

The words "genetic variability" tell us that *gene pools have lots of different alleles in them.* Every person has a set of genes that's different from that of every other individual (except that identical twins do have identical genes). What goes for people goes for every species on Earth. All individuals vary in their genes. For all populations of all species—there's genetic variability.

How does genetic variability happen? The answer is *random mutation. Mutation is the basis for genetic variability.*

- The phrase "genetic _____" refers to the fact that for any gene pertaining to any population of any species, the gene pool features a variety of alleles.

- The phrase "_____ variability" refers to the fact that within any population of any species, genotypes vary.

- Genetic variability [❑ is   ❑ is not] caused by a species' tendency to improve its relationship with its environment.

- Genetic variability [❑ is   ❑ is not] caused by a species' inability to adapt to existing environmental conditions.

- Genetic variability [❑ is   ❑ is not] attributable to random mutation.

- Genetic variability [❑ is   ❑ is not] a property of all populations.

GENETIC VARIABILITY AND EVOLUTION

When we say evolution we mean changing *gene pools*. Think about a population of frogs. Suppose half of the frogs have alleles for dark color and half have alleles for light color. Now imagine that an earthquake *separates* the frogs into two populations. When it's all over, some frogs are left in place 1 and others in place 2.

Places 1 and 2 differ. In place 1 there are predators that can see light-colored things, but *not* darkcolored things. In place 2 there are predators that can see light *and* dark colors. What's going to happen? In place 1, predators are going to prey happily on light-colored frogs. Lots of light-colored frogs will be eaten before they're even old enough to reproduce. But the predators won't have much luck with the dark frogs. Loads of dark frogs will live long enough to reproduce. They'll pass on the dark alleles to their offspring.

After a few frog generations the *frog gene pool in place 1 will change*. It won't be 50% light and 50% dark anymore. It might be 90% dark color and 10% light. Why? The light frogs are dying off—they get eaten before they reproduce. The dark frogs don't have that problem. They're hopping, swimming, jumping—and reproducing.

Meanwhile, what's happening in place 2? The predators in place 2 see light and dark frogs equally well. Things are as they were before the earthquake. The gene pool is 50% dark color and 50% light color.

That's Evolution: The Frogs in Place 1 Evolved

Evolution means that a population undergoes a change in the frequency of alleles in its gene pool. If there's a change in the frequency of alleles in the population's gene pool, the population evolves.

Our frogs in place 1 evolved because their environment changed. When it came to surviving in the new environment, dark frogs did better than light ones. They competed better. The frogs didn't know they were competing, but they *were*. They were competing to survive, and the dark ones were better competitors than the light ones. Why? Their dark color allowed them to escape their predators.

Individuals might compete for food, for water, or for all kinds of things in all kinds of ways for all kinds of reasons. *The better competitors are better at staying alive* and they have a better chance of reproducing. This is called natural selection. Natural selection just means that better competitors are better survivors and have a better chance of reproducing. Nature selects them to reproduce.

When organisms reproduce, they pass genes to their offspring. In each generation, more and more individuals resemble the better competitors; fewer and fewer resemble the poorer competitors. So the gene pool changes.

Now let's summarize. Evolution is a change in a population's gene pool. It happens because

- some individuals are better competitors than others. In the game of natural selection, they win.
- the winners survive, reproduce, and pass their genes on to their offspring.
- the offspring have genes like those of their parents. *Hence, the gene pool changes*. With each generation, it has more and more of the alleles that come from the better competitors.

Evolution and Species

Suppose two individuals—like a bumble bee and a dog—have such different genes that their gametes can't meet and form a new individual. We say the two individuals belong to two different species. When we say "different species," we mean individuals that can't produce offspring together.

Think again about our frogs, separated by an earthquake. Imagine that over time (hundreds of thousands of years perhaps)—the frogs in place 1 undergo so much change in their genes that they wouldn't be able to mate with the frogs in place 2, even if they'd been brought together with them again. We'd then say that evolution caused speciation. Because it creates two separate species, speciation is a form of divergent evolution. Divergent evolution happens when closely related species end up having different behaviors and traits. They used to have similar traits, but then the environment somehow pressured them to diverge: to show different traits from each other over many generations. If you see the term "divergent evolution" on the Biology Subject Test, think, "Two closely related species end up with different traits from each other, because they encountered different selection pressures from their environments." What kinds of environment might give rise to divergent evolution?

- Geographic barriers for populations of a species (such as mountains, land bridges, or rivers)
- Different habitats for populations of a land bridges species (for example, tree stumps vs. underground burrows)
- Different times of day, season, or year for breeding of populations of a species.

Review everything we've said about evolution. Then fill in the blanks and boxes.

- Evolution means a change in a population's _____ _____.

- Evolution [❑ **always** ❑ **sometimes**] results in the production of a new species.

- If a population is geographically divided, it [❑ **cannot** ❑ **may**] give rise to two separate species.

- Speciation [❑ **is** ❑ **is not**] a product of evolution.

- Speciation [❑ **increases** ❑ **decreases**] biology diversity.

- Divergent evolution can result from [❑ **only physical** ❑ **only behavioral** ❑ **both physical and behavioral**] selection pressures from the environment.

- In the course of divergent evolution, closely related species become [❑ **more** ❑ **less**] similar to each other with regard to behaviors and traits.

GETTING ORGANIZED: PHYLOGENY

When we try to organize people and their *addresses*, we could think like this:

Country • State • County • Town • Street • Street Number • Person

- The country has lots of states.
- Each state has lots of counties.
- Each county has lots of towns.
- Each town has lots of streets.
- Each street has lots of street numbers.
- At each street number, there might be several people.

Think about all the people living in one house. They have *a lot* in common. They live in the same house, on the same street, in the same town, in the same county, in the same state, in the same country.

Think about all people living on the same *street*. They have a lot in common too: street, town, county, state, and country. They don't have as much in common as do the people living in the same *house*, but they do have a lot in common.

How about people living in the same town? They have a few things in common too: town, county, state, and country. They have *less* in common than do the people living on the same street, and less still than do the people living in the same house.

So think about it:

- People living in the same house have more in common than do people living simply on the same street;
- People living on the same street have more in common than do people living simply in the same town;
- People living in the same town have more in common than do people living simply in the same county;
- People living in the same county have more in common than do people living simply in the same state;
- People living in the same state have more in common than do all people living simply in this country.

COUNTRY • STATE • COUNTY • TOWN • STREET • STREET NUMBER • PERSON

LESS IN COMMON → → → → MORE IN COMMON

The Same Goes for Species

There are lots of species on the Earth. Biologists assign them "addresses" using this arrangement:

KINGDOM • PHYLUM • CLASS • ORDER • FAMILY • GENUS • SPECIES

LESS IN COMMON → → → → MORE IN COMMON

Here's how to remember that:

King   **P**hillip   **C**ame   **O**ver   **F**rom   **G**ermany—**S**o?

Kingdom • **P**hylum • **C**lass • **O**rder • **F**amily • **G**enus • **S**pecies

Look at the organizational scheme and realize:

- Each kingdom is made up of many phyla.
- Each phylum is made up of many classes.
- Each class is made up of many orders.

- Each order is made up of many families.
- Each family is made up of many genuses.
- Each genus is made up of many species.

Realize also that:
- Organisms of the same species have more in common than do organisms belonging simply to the same genus.
- Organisms of the same genus have more in common than do organisms belonging simply to the same family.
- Organisms of the same family have more in common than do organisms belonging simply to the same order.
- Organisms of the same order have more in common than do organisms belonging simply to the same class.
- Organisms of the same class have more in common than do organisms belonging simply to the same phylum.
- Organisms of the same phylum have more in common than do organisms belonging simply to the same kingdom.

What do we actually mean when we say that organisms have things in common? We mean that different organisms end up having common ancestors, and they share some traits from those ancestors of long ago. Scientists classify organisms based on their evolutionary relationships (phylogeny). Taxonomy is the fancy term given to the science of classification.

Remember this man's name: Carolus Linnaeus. Carolus Linnaeus came up with what is known as the modern system of classification, called the binomial system. The binomial system of classification is based on a two-part name for each organism. The first part is the organism's *Genus*, and the second part is the organism's *species*. The Genus is capitalized, but the species is not. Your pet dog is *Canis familiars*; a wolf is *Canis lupus*. Sugar maple is *Acer saccharum*; humans are *Homo sapiens*.

Got it? Let's see. Fill in the blanks and boxes appropriately.

- The science of classification is called [❑ **taxidermy** ❑ **taxonomy**].

- The conventional ordering of phylogeny is kingdom, _____, _____, _____, _____, _____, _____.

- The binomial system of nomenclature was conceived by [❑ **Charles Darwin** ❑ **Carolus Linnaeus**].

- The members of a kingdom [❑ **do** ❑ **do not**] have more in common than do the members of an order.

- The members of an order [❑ **do** ❑ **do not**] have more in common than do the members of a class.

WHO EVOLVED BEFORE WHOM

The test writers might want you to know the order in which certain modern-day animals appeared on the Earth. Life began in the water, so simple aquatic animals such as sponges and hydra came first. They're two-layered organisms that do everything (eat, excrete, and breathe) through diffusion. Then came the worms (flatworms, roundworms, flukes and tapeworms, which are parasites, and segmented worms). They're three-layered organisms with well-formed organs. The most important worm as far as the test writers are concerned is the earthworm. The earthworm has a closed circulatory system, breathes through its skin and excretes wastes with special organs called nephridia.

Some animals have spiny exoskeletons, like a sea urchin. Other animals have soft bodies and hard shells, like a snail. Arthropods have jointed appendages, an exoskeleton and a segmented body. The most complex animals are the chordates, which include fish, amphibians, reptiles, birds and mammals. How do you make sense of all this? Study this handy classification chart. It may look imposing now, but the more you look it over, the easier it will get to remember who's who.

1. *Kingdom: Monera* — *Monera are prokaryotic; they lack a nuclear membrane and all organelles. They have circular DNA.*

    Phylum: Bacteria — Most bacteria are heterotrophic. They can form mutualistic relationships (e.g. *E. coli*) or they are pathogens or decomposers.

    Phylum: Cyanobacteria — Cyanobacteria (blue-green algae) contain chlorophyll and can photosynthesize.

2. *Kingdom: Protista* — *Protists are eukaryotic; they contain organelles and a nucleus.*

    Phylum: Protozoa — Protozoa are unicellular and animal-like. Examples are the amoeba and paramecium.

    Phylum: Slime mold — Slime molds resemble an overgrown amoeba. They contain protoplasm and many nuclei.

    Phylum: Algae — Algae are unicellular and live in colonies. They possess a cell wall and chloroplasts, making them photosynthetic. Examples are kelp and diatoms.

3. *Kingdom: Fungi* — *Fungi are eukaryotic; they have a filamentous structure and many nuclei. They lack a digestive system and are decomposers. Examples are mushrooms and yeast.*

4. *Kingdom: Plants* — *Plants are eukaryotic, multicellular, and photosynthetic.*

    Phylum: Bryophytes — Bryophytes do NOT have xylem and phloem, nor do they have true stems, leaves, or roots. Examples are mosses and liverworts.

	Phylum: Tracheophytes	Tracheophytes DO have xylem and phloem; they also have true roots, stems, and leaves. Examples are ferns, oaks, and tulips.
	Subphylum: Gymnosperm	Gymnosperms have naked seeds that are located in the cones of evergreens. An example of a gymnosperm is a pine tree.
	Subphylum: Angiosperm	Angiosperms have enclosed seeds located in a fruit or nut. Examples are lima beans and flowers.
	Class: Monocots	Monocots have one cotyledon, leaves that have parallel veins, and flower parts that come in multiples of three.
	Class: Dicots	Dicots have two cotyledons and leaves that have net-like veins.
5.	*Kingdom: Animalia*	*Animalia are eukaryotic, multicellular, and heterotrophic.*
	Phylum: Coelenterates	Coelenterates possess two cell layers and a sack-like digestive system. Examples are hydra and jellyfish.
	Phylum: Annelids	Annelids are segmented worms. They possess a closed circulation, a mouth and anus, and excrete waste through nephridia. An example is the earthworm.
	Phylum: Echinoderms	Echinoderms exhibit radial symmetry and spiny exoskeletons. An example is the sea urchin.
	Phylum: Mollusks	Mollusks have soft bodies and hard outer shells. An example is a snail.
	Phylum: Arthropods	Arthropods have jointed appendages, an exoskeleton containing chitin, and a segmented body with head, thorax, and abdomen. Arthropods include crustaceans (e.g., crabs), insects (e.g., moths), and arachnida (e.g., spiders).
	Phylum: Chordates	Chordates possess a hollow notochord, a dorsal nerve cord, and a tail. Chordates include fish, amphibians, reptiles, birds (aves), and mammals.

In terms of chordates, fish came first. Then there came some animals that could hang around water *and* land; those are the amphibians (who possess a three-chambered heart). Gradually, there appeared a bunch of species that preferred land to water, but they didn't mind jumping in and out of water now and then. Those are the reptiles (who also have a three-chambered heart). After the reptiles came the birds (who have a four-chambered heart) and after the birds came the mammals.

Since we're talking about animals you might find it easy to remember all of this by thinking of the word "FARM"—with a "B" stuck into it: "FARBM."

F: Fish

A: Amphibians

R: Reptiles

B: Birds

M: Mammals

Fill in the blanks and boxes appropriately.

- With respect to the evolution of life forms, reptiles appeared [❏ **before** ❏ **after**] amphibians.

- With respect to the evolution of life forms, fish appeared [❏ **before** ❏ **after**] birds, and birds appeared [❏ **before** ❏ **after**] mammals.

- The order in which life forms appeared on Earth is
  (1) fish, (2) _____, (3) _____,
  (4) _____, (5) _____.

EVERY SPECIES HAS AN ECOLOGICAL NICHE

Niche refers to an organism's relationship with the whole system in which it lives. It has to do with the way an organism meets its needs and the way it helps or hurts other organisms in the system. Niche refers to all of the elements of an organism's lifestyle.

Niche isn't something you can see or touch; it's an abstract idea. When we say niche we mean that every species, plant or animal, has its own *place* in the community. It uses one or more plants or animals for food. It may serve as food for some other organism. It might grab nitrogen out of the air and give it to plants. It might eat zebras and leave scraps for buzzards. It might do all kinds of things that help itself, meanwhile helping or hurting others.

Fill in the blanks and boxes appropriately.

- The term "niche" [❏ **does** ❏ **does not**] refer solely to the physical habitat in which an organism lives.

- The term "niche" [❏ **does** ❏ **does not**] refer, abstractly, to all aspects of an organism's relationship to the biological system in which it lives.

- The term "niche" is fairly synonymous with [❏ **habitat** ❏ **food chain** ❏ **lifestyle**].

FOOD CHAINS

You need to know something about the way organisms go about the happy business of eating one another. That means you must know about food chains:

1. The food chain starts with organisms we call "producers." They're photosynthetic plants—autotrophs. They take energy from the sun and convert it to carbohydrates.

2. Who eats these producers? Organisms we call primary consumers—the herbivores—animals that eat plants.

3. Who eats the primary consumers? Organisms we call secondary consumers—the carnivores—animals that eat other animals.

Don't Forget Decomposers

One way or another, all organisms die. As soon as they do, little fellows called decomposers come along and eat their dead bodies. Who are they? Bacteria and fungi who live off the dead remains of animals. Bacteria and fungi are decomposers. Remember:

- **producers** (autotrophs)
- **primary consumers** (herbivores) eat **producers** (photosynthetic plants)
- **secondary consumers** (heterotrophs and carnivores) eat primary consumers.
- tertiary consumers eat secondary consumers
- **decomposers** eat the dead.

Think of the food chain as a pyramid.

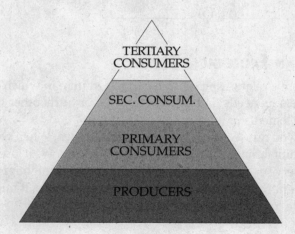

There are more species and individuals at the bottom of the food chain than at the top.

Tertiary Consumers→ Secondary Consumers →Primary Consumers→ Producers

FEWER→ → → → MORE

There are more producers than primary consumers, and more primary consumers than secondary consumers.

What about decomposers? There are lots and lots of them. Don't worry about how many.

There is also more *energy* available at the bottom of the food chain than at the top. Each tier (level) of the food chain going *up* has less energy to it than the tier below it.

Think about a meadow. Grass grows there. Field mice live there, and eat the grass. A hawk cruises the meadow for mice. A fox also roams the meadow in search of mice.

Let's sum up what we know about this particular food chain:

- ◆ The grass is the producer.
- ◆ The mice are the primary consumers.
- ◆ The fox and the hawk are the secondary consumers.
- ◆ Mice will exist in greater numbers than foxes or hawks.
- ◆ More energy is available to the mice than to foxes or hawks.

- In a food chain involving grass, mice, snakes, and hawks, the mice are _____ _____ and the hawks are tertiary _____.

- In a food chain consisting of grass, grasshoppers, frogs, and bass, grasshoppers are [❑ **more** ❑ **less**] numerous than frogs, and frogs are [❑ **more** ❑ **less**] numerous than bass.

- Heterotroph is synonymous with _____ and *autotroph* is synonymous with _____.

- The energy available to an organism [❑ **increases** ❑ **decreases**] with each successive feeding level of a food chain, beginning with producers.

- Decomposers include _____ and _____.

- Primary consumers are called _____vores and secondary consumers are called _____vores.

- An herbivore is a(n) [❑ **autotroph** ❑ **heterotroph**].

LET'S TALK ABOUT A THING CALLED ECOLOGICAL SUCCESSION

We've been talking about biological communities and the way their inhabitants live together (and the way in which they eat one another). For the Biology Subject Test you should also know that communities of organisms undergo change over time—just like residential communities of people.

Think first about human beings and an uninhabited area outside a major city. Suppose that in 1950 a few pioneers decide to build houses in the region and move in. Next, a couple of stores appear. More people become interested in the locality and they move in too. The region develops slowly into a rural suburb. By 1960 it spawns some housing developments and a few little shopping strips. By 1965 roads widen, traffic increases, and lots of gas stations appear. The process continues—gradually. By 1994 the whole place is quite different from what it was in 1950. It boasts big shopping malls, big roads, lots of traffic, schools, businesses, apartment houses, and lots of people. It starts to look like a city.

When that process of gradual change takes place in an *ecological* community we say ecological succession. Think, for instance, about a rock with no living things on it. Often the first resident to move in to the area is a thing called lichen—lichen are the pioneers, and biologists actually call them pioneer organisms because they're the first to start living in a previously uninhabited area. When you think lichen think pioneer organism.

As the lichen flourish, they give off a bunch of substances that corrode the rock. The rock represents an altered environment and it becomes attractive to other organisms like mosses and herbs. The mosses and herbs further affect the environment to make it suitable for shrubs and grasses. Trees move in next—first pine trees and then deciduous trees, like oaks, beeches, and maple. The rock—formerly lifeless—gave rise to an ecological community; it underwent succession which begins, always, with some pioneer organism.

Let's run through that succession again, because it may show up on the Biology Subject Test:

rock → lichen → mosses and herbs → grasses and shrubs
→ pine trees → deciduous trees

We said that the lichen in this succession is the pioneer organism. Does anyone else get a special label? The answer is yes. The deciduous trees—the oaks, beeches, and maples—are called the climax community. Once the succession has progressed as far as it can go, what you have left is the climax community. Unlike all of the plant communities that came before it (the lichen, the mosses and herbs, the grasses and shrubs, and the pine trees), the climax community is here to stay. Biologists call the climax community a stable community; it doesn't change. Lots of things still happen *within* the climax community—plants and animals come and go, live and die, storms may hit it—but the essential nature of the climax community—that it's made up of deciduous trees and has a characteristic fauna associated with it—stays the same.

Don't Confuse Ecological Succession with Evolution

Ecological succession is not about evolution (a whole different ball game). Ecological succession has to do with continuous changes that take place in a community. There are predictable stages to it, as we just saw with our rock-lichen-grass-shrubs-trees example. Another big tip-off that you're dealing with an ecological succession is the time frame involved: it's a whole lot shorter than the time frame required for evolution to take place. When you think ecological succession, think: predictable stages of plant communities usually over a period of decades.

Fill in the blanks and boxes appropriately.

- The phrase "ecological succession" refers to the development or alteration of biological _____.

- The organism that first appears as founder of a biological community is called a _____ organism.

- When a plant community begins on a barren rock, the pioneer organism is most often [❏ **lichen** ❏ **moss**].

- Each plant community that inhabits an area contributes to the process of ecological succession by modifying the environment, making it [❏ **more** ❏ **less**] favorable for itself, and [❏ **more** ❏ **less**] favorable for another plant community.

- In the ecological succession depicted below, the climax community is composed of _____ _____.

    pond → cattails → mosses and sedges → shrubs
    → pine trees → deciduous trees

- In an ecological succession, each new plant community in an area [❏ **coexists with** ❏ **replaces**] the previous plant communities.

- In the ecological succession depicted below, impermanent plant communities include _____ _____, and _____ _____.

    area → pioneer organism → plant community A → plant community B
    → plant community C → climax community

BIOLOGICAL COMMUNITIES DEVELOP: BIOMES

SAT II Subject Test writers might ask about biomes. Biomes are great big areas classified mostly by plant life and climate. Organisms of all types somehow live together in the biomes (loving, hating, eating, and killing). Biomes have names. Here are the five you should know about:

Biome 1—The tundra: A tundra is a big area whose soil is frozen all year long. We find a tundra in the northernmost parts of North America, Europe, and Asia. There are few trees on the tundra. There are some mammals like reindeer, wolves, foxes, grizzlies, and Kodiac bears. Plant life includes, primarily, lichens, mosses, grasses, and wild flowers.

Biome 2—The taiga: A little bit south of the tundra is a biome we call taiga. The main thing to remember about the taiga is that it's packed with evergreen trees. Evergreen trees are also called conifers. The taiga's fauna (animal life) include moose, caribou, beavers, and grizzly bears.

Biome 3—Deciduous forests: South of the taiga we have the deciduous forests, where there's lots of rain and a variety of plants and animals. The climate has distinct hot/cold seasons. There are many more life forms in the deciduous forests than there are in the tundra or taiga. Examples of fauna are deer, skunk, beavers, raccoon, foxes, black bears, and squirrels. Examples of flora (plant life) are maple trees, elm trees, oak trees, and chestnut trees.

Biome 4—Tropical rainforest: The tropical rain forests are south of the deciduous forests. They have loads of rainfall. Tropical rain forests have *tons and tons* of organisms; in other words, they have rich and diverse plant life. On the other hand, the soil of tropical rain forests is typically nutrient-poor, because the nutrients are all tied up in the plant life there.

Biome 5—Desert: Deserts are dry. There's hardly any rainfall. The plants and animals that live there can do without water for a long time. If a plant can live without water for a long time, we call it succulent. That means it can store lots of water for long periods. Examples of desert plants are cacti and aloe vera. Examples of desert animals are the kangaroo rat and jackrabbit.

Review what we've said about biomes. Then fill in the blanks and boxes appropriately.

- The biome with frozen soil year-round and the fewest number of trees is the _____.

- The biome with the least rainfall and a population of succulent plants is the _____.

- The biome with the greatest number of organisms is the _____ _____ forest.

- The biome characterized by a large number of conifers (evergreen trees) is the _____.

- The tropical rain forest has [❏ more ❏ fewer] life forms than the tundra.

- The taiga [❏ is ❏ is not] rich in conifers.

- Deer, black bears, and raccoon are found in the [❏ tundra ❏ rainforest ❏ temperate deciduous forest], while caribou, beaver, and moose are found in the [❏ tundra ❏ rain forest ❏ temperate deciduous forest].

- The primary plant forms of the [❏ taiga ❏ tundra ❏ desert] are mosses, lichens, and wildflowers.

- The primary flora of the desert are the succulents, such as _____ and _____ _____.

6

The Princeton Review SAT II: Biology Regular Subject Test

BIOLOGY SUBJECT TEST

SECTION 1

Your responses to the Biology Subject Test questions must be filled in on Section One of your answer sheet (the box on the front of the answer sheet). Marks on any other section will not be counted toward your Biology Subject Test score.

When your supervisor gives the signal, turn the page and begin the Biology Subject Test. There are 100 numbered ovals on the answer sheet and 95 questions in the Biology Subject Test. Therefore, use only ovals 1 to 95 for recording your answers.

BIOLOGY TEST

Part A

Directions: Each of the questions or incomplete statements below is followed by five suggested answers or completions. Select the one that is best in each case and then fill in the corresponding oval on the answer sheet.

1. Which of the following accurately states the principle(s) of the cell theory?

    I. All organisms are composed of cells.
    II. All cells arise from preexisting cells.
    III. Cells are the basic unit of biological function.

    (A) I only
    (B) II only
    (C) I and II only
    (D) II and III only
    (E) I, II, and III

2. The plasma membrane is a semipermeable organelle composed chiefly of

    (A) lipids and sugars
    (B) sugars and proteins
    (C) proteins and lipids
    (D) lipids and carbohydrates
    (E) carbohydrates and proteins

3. The process of anaerobic respiration can lead to the formation of

    (A) lactic acid or alcohol
    (B) lactic acid or amino acids
    (C) carbon monoxide and water
    (D) oxygen and carbon
    (E) ammonia or alcohol

4. Which of the following is NOT found in a molecule of DNA?

    (A) Adenine
    (B) Deoxyribose
    (C) Phosphorus
    (D) Uracil
    (E) Thymine

5. According to the heterotroph hypothesis, the atmosphere of the Earth before the beginning of life contained all of the following gases EXCEPT

    (A) hydrogen
    (B) water
    (C) ammonia
    (D) methane
    (E) oxygen

6. The process in which a dipeptide is broken into two amino acids is called

    (A) hydrolysis
    (B) dehydration synthesis
    (C) peptide bonding
    (D) hydrogen bonding
    (E) transcription

7. The amoeba, paramecium, and euglena are classified in the same phylum because all

    (A) have cell walls
    (B) reproduce by spore formation
    (C) are photosynthetic
    (D) have external digestive systems
    (E) are unicellular

8. Plant cells differ from animal cells in that plant cells

    (A) perform respiration
    (B) perform protein synthesis
    (C) contain a cell membrane
    (D) contain chloroplasts
    (E) contain vacuoles

GO ON TO THE NEXT PAGE

BIOLOGY – Continued

9. Which of the following accurately describes a principal difference between the human nervous and endocrine systems?

    (A) Endocrine hormones travel through the blood, and nervous signals travel through neurons.
    (B) Endocrine organs secrete neurotransmitters and nerves secrete hormones.
    (C) The endocrine system relies on chemical transmission, and the nervous system does not.
    (D) The endocrine system involves myelinated tissues, and the nervous system does not.
    (E) The endocrine system is involved in homeostasis, and the nervous system is not.

10. Which of the following is involved in the transport of food within green plants?

    (A) Tracheids
    (B) Sieve-tube cells
    (C) Cambium
    (D) Stomates
    (E) Guard cells

11. Which of the following is true concerning producers?

    (A) They are heterotrophic.
    (B) They represent the smallest single component of the Earth's biomass.
    (C) They derive their energy solely from organic compounds.
    (D) They derive their energy directly from solar energy.
    (E) They are carnivorous.

12. An mRNA molecule carries the codon sequence adenine–uracil–cytosine. Which of the following represents the complementary triplet on a tRNA?

    (A) Thymine-adenine-guanine
    (B) Guanine-adenine-thymine
    (C) Uracil-adenine-guanine
    (D) Uracil-uracil-guanine
    (E) Uracil-guanine-cytosine

13. The basic principles of genetics were established in the nineteenth century by

    (A) Jean Lamarck
    (B) Louis Pasteur
    (C) Charles Darwin
    (D) James Watson
    (E) Gregor Mendel

14. The roots of plants serve which of the following functions?

    I. Anchorage
    II. Nutrient absorption
    III. Photosynthesis

    (A) I only
    (B) II only
    (C) I and II only
    (D) II and III only
    (E) I, II, and III

GO ON TO THE NEXT PAGE

BIOLOGY – Continued

15. Among the following, the phylogenetic group whose members show the greatest degree of similarity is

    (A) phylum
    (B) class
    (C) order
    (D) species
    (E) family

16. A principal difference between heterotrophs and autotrophs is that

    (A) heterotrophs produce their own nutrients and autotrophs do not
    (B) heterotrophs rely on other organisms for nutrition and autotrophs do not
    (C) heterotrophs feed only on dead organisms and autotrophs feed on living organisms
    (D) heterotrophs perform photosynthesis and autotrophs do not
    (E) heterotrophs require oxygen and autotrophs do not

17. A yeast cell may undergo reproduction asexually by a process known as

    (A) sporulation
    (B) budding
    (C) vegetative propagation
    (D) binary fission
    (E) regeneration

18. Which of the following is more primitive than a bird but more advanced than an amphibian?

    (A) Amoeba
    (B) Fish
    (C) Reptile
    (D) Mammal
    (E) Slime mold

19. A principal difference between mitosis and meiosis is that

    (A) mitosis involves physical division and meiosis does not
    (B) mitosis involves replication of chromosomes and meiosis does not
    (C) mitosis occurs only in sex cells and meiosis occurs only in somatic cells
    (D) mitosis occurs only in males and meiosis occurs only in females
    (E) mitosis gives rise to diploid daughter cells and meiosis gives rise to haploid daughter cells

20. Which of the following correctly characterizes the process of evolution?

    I. Populations evolve, not individuals
    II. Mutations tend to produce genetic variability
    III. It is a change in a population's allele frequencies

    (A) I only
    (B) II only
    (C) I and II only
    (D) II and III only
    (E) I, II, and III

21. Which of the following is most properly associated with the term *pioneer organism*?

    (A) Oak tree
    (B) Lichen
    (C) Grass
    (D) Conifer
    (E) Shrub

BIOLOGY – Continued

22. A population of a single species of birds is divided by natural disaster and the two populations are subjected to different environmental conditions. After a period of 6,000,000 years it is most likely that the descendants of the two populations, if brought together, would

    (A) occupy the same niche
    (B) have identical mating seasons
    (C) be in competition for identical resources
    (D) be unable to defend themselves against predators
    (E) be unable to mate together

23. A principal effect of insulin produced by the pancreas is to

    (A) stimulate metabolism
    (B) increase blood glucose levels
    (C) decrease blood glucose levels
    (D) regulate secondary sex characteristics
    (E) promote bodily growth

24. The organ that secretes the main digestive juices for complete digestion is known as the

    (A) mouth
    (B) esophagus
    (C) pancreas
    (D) small intestine
    (E) large intestine

25. All of the following hormones are involved in the menstrual cycle EXCEPT

    (A) progesterone
    (B) estrogen
    (C) testosterone
    (D) follicle-stimulating hormone (FSH)
    (E) luteinizing hormone (LH)

26. The musculoskeletal system arises from which of the following embryonic germ layers?

    (A) Endoderm
    (B) Mesoderm
    (C) Ectoderm
    (D) Epidermis
    (E) Epididymus

27. The exchange of corresponding chromatid segments between homologous chromosomes occurs during

    (A) Prophase I
    (B) Metaphase I
    (C) Anaphase I
    (D) Prophase II
    (E) Metaphase II

28. In a marathon runner, painful muscle fatigue may be caused by the products of

    (A) carbon fixation
    (B) aerobic respiration
    (C) anaerobic respiration
    (D) lipolysis
    (E) proteolysis

GO ON TO THE NEXT PAGE

BIOLOGY — Continued

29. Which of the following equations represents the process of photosynthesis?

    (A) Carbon dioxide + water → glucose + oxygen + water
    (B) Glucose + oxygen + water → carbon dioxide + water
    (C) Glucose + water → carbon dioxide + water + oxygen
    (D) Ammonia + water → glucose + oxygen + water
    (E) Carbon + water → glucose + carbon dioxide + water

30. In chick embryos, the extraembryonic membrane that functions in excretion is called the

    (A) allantois
    (B) chorion
    (C) amnion
    (D) yolk sac
    (E) egg shell

31. In humans, gas exchange occurs principally within

    (A) the alveoli
    (B) the trachea
    (C) the nephron
    (D) the lymph glands
    (E) the nostrils

32. Which of the following exhibit an exoskeleton?

    (A) Insects
    (B) Birds
    (C) Fish
    (D) Worms
    (E) Mammals

33. For a guinea pig, black coat (B) is dominant over white coat (b). If two guinea pigs mate and produce 75% offspring with black coats and 25% offspring with white coats, then the genotypes of the parent organisms are most likely

    (A) BB × BB
    (B) Bb × BB
    (C) BB × bb
    (D) Bb × Bb
    (E) bb × Bb

34. All of the following represent correctly paired organic compounds and the products of their decomposition EXCEPT

    (A) protein → amino acids
    (B) glucose → lactic acid
    (C) lipids → fatty acids
    (D) cellulose → carbohydrates
    (E) starch → simple sugars

GO ON TO THE NEXT PAGE

BIOLOGY – Continued

35. In humans, which of the following correctly represents the sequence of embryological development?

    (A) fertilization → zygote → gametes → embryo → cleavage
    (B) fertilization → gametes → cleavage → zygote → embryo
    (C) gametes → fertilization → zygote → cleavage → embryo
    (D) cleavage → gametes → embryo → fertilization → zygote
    (E) zygote → fertilization → embryo → gametes → cleavage

36. Which of the following represents the approximate ratio of carbon to hydrogen in a molecule of sucrose?

    (A) 1 : 1
    (B) 1 : 2
    (C) 1 : 3
    (D) 2 : 1
    (E) 3 : 1

37. Parathormone, which is produced by the parathyroids, regulates the metabolism of

    (A) calcium
    (B) glucose
    (C) potassium
    (D) iodine
    (E) zinc

38. All of the following compounds are organic EXCEPT

    (A) glycogen
    (B) fats
    (C) maltose
    (D) water
    (E) starch

39. Which of the following graphs best describes the relationship between reaction rate as a function of the amount of substrate if the amount of enzyme is limited?

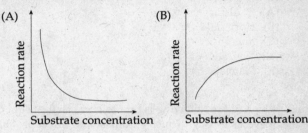

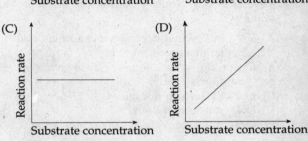

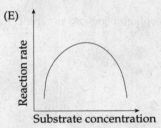

GO ON TO THE NEXT PAGE

BIOLOGY — Continued

40. All of the following compounds are carbohydrates EXCEPT

    (A) cellulose
    (B) starch
    (C) glucose
    (D) glycogen
    (E) lipid

41. Which of the following are NOT found circulating in human blood?

    (A) Erythrocytes
    (B) Hormones
    (C) Lymphocytes
    (D) Neurons
    (E) Platelets

42. Which of the following is first to occur in the sequence of events through which chromosomes direct protein synthesis?

    (A) DNA nucleotide sequences cause amino acids to gather in a specific order.
    (B) RNA codons cause amino acids to gather in appropriate order.
    (C) mRNA travels to ribosomes.
    (D) DNA generates mRNA.
    (E) tRNA anticodon attaches to mRNA codon.

43. In mammalian species, epinephrine and norepinephrine are released by the

    (A) heart
    (B) muscles
    (C) adrenal medulla
    (D) thyroid
    (E) pancreas

44. Which of the following correctly orders the events associated with the discharge of a neuron?

    (A) polarization → depolarization → action potential → repolarization
    (B) repolarization → action potential → polarization → depolarization
    (C) action potential → polarization → repolarization → depolarization
    (D) depolarization → repolarization → action potential → polarization
    (E) depolarization → polarization → action potential → repolarization

45. Which of the following structures are found among bacteria?

    I. Plasma membrane
    II. Cell wall
    III. DNA

    (A) I only
    (B) II only
    (C) I and II only
    (D) II and III only
    (E) I, II, and III

BIOLOGY — Continued

Questions 46-50 refer to the enzymatic breakdown of maltose as shown schematically below.

46. Which structure represents a substrate not yet attached to its enzyme's active site?

    (A) 1
    (B) 2
    (C) 4
    (D) 5
    (E) 7

47. Which structure is identical to structure 1?

    (A) 3
    (B) 4
    (C) 5
    (D) 6
    (E) 7

48. If structure 1 were replaced with another substance the reaction would not occur due to

    (A) enzyme specificity
    (B) denaturation
    (C) dehydration synthesis
    (D) fermentation
    (E) peptide bond formation

49. According to the schematic diagram presented above, the enzymatic breakdown of maltose requires the addition of

    (A) oxygen
    (B) hydrogen
    (C) water
    (D) carbon dioxide
    (E) adenosine triphosphate

50. Among the following, which term best describes the enzyme that catalyzes the breakdown of maltose?

    (A) Polypeptide chain
    (B) Amino acid
    (C) Carbohydrate
    (D) Starch
    (E) Nucleic acid

GO ON TO THE NEXT PAGE

BIOLOGY – *Continued*

Questions 51-54 refer to the schematic diagram of a human with endocrine organs shown in place:

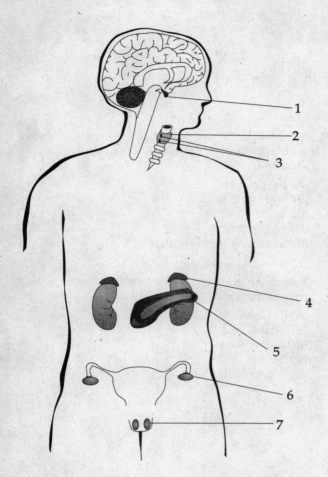

Male and Female Organs are shown

51. Which gland secretes a hormone that reduces blood levels of glucose?

    (A) 1
    (B) 2
    (C) 4
    (D) 5
    (E) 7

52. Which organ secretes thyroxine?

    (A) 2
    (B) 3
    (C) 4
    (D) 5
    (E) 6

53. Which gland has an anterior portion that secretes adrenocorticotropic hormone (ACTH) and a posterior portion that secretes antidiuretic hormone (ADH)?

    (A) 1
    (B) 2
    (C) 4
    (D) 5
    (E) 7

54. Which gland has an outer cortex that secretes glucocorticoids and mineralocorticoids and an inner portion that secretes epinephrine and norepinephrine?

    (A) 2
    (B) 4
    (C) 5
    (D) 6
    (E) 7

BIOLOGY — Continued

Part B

Directions: Each set of lettered choices below refers to the numbered statements immediately following it. Select the one lettered choice that best fits each statement and then fill in the corresponding oval on the answer sheet. A choice may be used once, more than once, or not at all in each set.

Questions 55-58 refer to the following diagram of a portion of the human female anatomy.

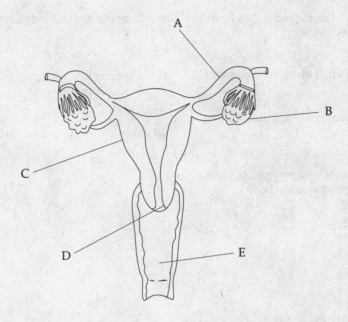

55. Location at which maturation of gametes occurs

56. Organ in which embryo is implanted

57. Organ in which estrogen is produced

58. Site of fertilization of ovum

BIOLOGY – *Continued*

Questions 59-62

    (A) Right atrium
    (B) Aorta
    (C) Pulmonary artery
    (D) Right ventricle
    (E) Left atrium

59. Artery through which oxygenated blood leaves heart to be carried throughout body

60. Artery that carries deoxygenated blood from the heart

61. Chamber in the heart that receives blood directly from pulmonary circulation

62. Chamber in the heart that receives deoxygenated blood returning from locations throughout the body

Questions 63-65

    (A) Tundra
    (B) Tropical forest
    (C) Taiga
    (D) Desert
    (E) Temperate deciduous forest

63. Fauna includes moose and black bear

64. Flora consists mainly of lichen, mosses, and grasses

65. Contains drought-resistant shrubs and succulent plants

Questions 66-68

    (A) Ribosomes
    (B) Mitochondria
    (C) Nucleus
    (D) Lysosomes
    (E) Cell membrane

66. Contains the enzymes of the Krebs cycle

67. Contains hydrolytic enzymes that participate in intracellular digestion

68. Site of DNA transcription

GO ON TO THE NEXT PAGE

BIOLOGY — Continued

Questions 69-71 refer to the following diagram of a flowering plant.

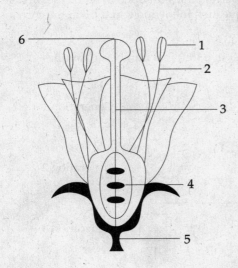

69. Which structure contains female monoploid nuclei?

    (A) 1
    (B) 2
    (C) 3
    (D) 4
    (E) 5

70. Pollen grains are produced by

    (A) 1
    (B) 3
    (C) 4
    (D) 5
    (E) 6

71. Pollen grains germinate on structure

    (A) 1
    (B) 2
    (C) 4
    (D) 5
    (E) 6

Questions 72-74 refer to the following diagram of the human digestive system.

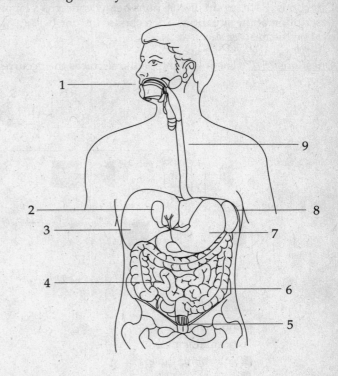

72. Bile is stored in

    (A) 2
    (B) 3
    (C) 4
    (D) 7
    (E) 8

73. Insulin is produced in

    (A) 1
    (B) 3
    (C) 4
    (D) 7
    (E) 8

74. Amylase is first secreted and carbohydrate digestion begins in

    (A) 1
    (B) 2
    (C) 4
    (D) 6
    (E) 9

GO ON TO THE NEXT PAGE

BIOLOGY – Continued

Part C

Directions: Each group of questions below concerns a laboratory or experimental situation. In each case, first study the description of the situation. Then choose the one best answer to each question following it and fill in the corresponding oval on the answer sheet.

Questions 75-77: The following pedigree traces the occurrence of a recessive trait in a number of families.

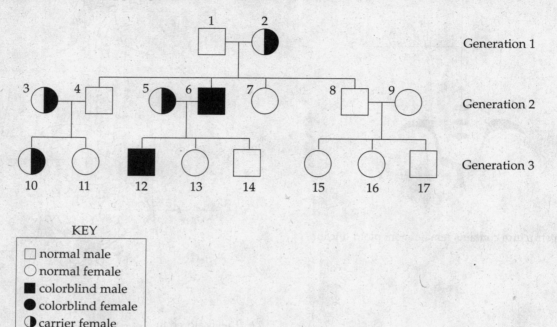

75. Among the following, what would constitute evidence that the trait is sex-linked?

    (A) The trait never occurs in females.
    (B) The trait never occurs in males.
    (C) The trait never passes from a male parent to a female child.
    (D) The trait never passes from a male parent to a male child.
    (E) The trait never passes from a female parent to a male child.

76. The pedigree shown above indicates that if individual 10 bears a child with a colorblind male, the probability that the child will be colorblind is

    (A) 0%
    (B) 25%
    (C) 50%
    (D) 75%
    (E) 100%

77. According to the pedigree shown above, which of the following groups of individuals is heterozygous for the colorblind trait?

    (A) Individuals 2, 3, 5, and 10
    (B) Individuals 2, 5, 8, 10, and 12
    (C) Individuals 3, 5, 9, 11, and 13
    (D) Individuals 2, 3, 5, 6, and 16
    (E) Individuals 2, 3, 4, 10, and 14

BIOLOGY — Continued

Questions 78-80

An experimenter wishes to study the manner in which newly hatched birds acquire attachments to their mothers. Six newly hatched ducklings were selected. At the moment of birth, ducklings 1, 2, and 3 were removed from the birth site so that they had no exposure to their mother. Ducklings 4, 5, and 6 remained with the mother.

A balloon was then shown to ducklings 1, 2, and 3 and simultaneously each duckling was exposed to a recording of a mother duck call. It was found over subsequent hours and days that ducklings 4, 5, and 6 imitated their mother's behavior and consistently followed her about. It was also found that ducklings 1, 2, and 3 exhibited behavior similar to that of the balloon they had seen and attempted to follow all balloons to which they were subsequently exposed.

78. The process that occurred in relation to ducklings 1, 2, and 3 is termed

    (A) innate behavior
    (B) reflex behavior
    (C) instinct
    (D) conditioned response
    (E) imprinting

79. Ducklings 4, 5, and 6 would best be described as

    (A) experimental models
    (B) controls
    (C) dependent variables
    (D) independent variables
    (E) experimental sets

80. If instead of exposing ducklings 1, 2, and 3 to a balloon and a recorded duck call, the experimenter had exposed them to a cat and a recorded duck call, which of the following would most likely occur in relation to ducklings 1, 2, and 3?

    (A) They would fear other ducklings.
    (B) They would lose their ability to reproduce.
    (C) They would acquire an ability to make sounds resembling those of a cat.
    (D) They would regard the cat as their mother.
    (E) They would believe themselves to be cats.

BIOLOGY – Continued

Questions 81-83 refer to the data presented in the table below, resulting from an experiment concerning the rate of growth in a bacterial population over a 25-hour period.

Column I	Column II
Time (hrs)	Number of bacteria
0	750
5	9,000
10	44,000
15	35,000
20	11,000
25	6,000

81. Which of the following graphs best represents the results set forth in the table above?

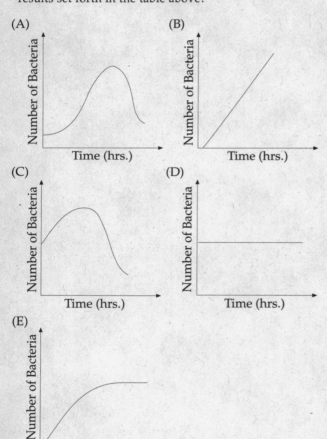

82. The most probable explanation for the numerical decline observed for the last three entries in column 2 is that the bacteria

    (A) exhausted their ability to manufacture respiratory enzymes
    (B) exhausted their capacity to undergo binary fission
    (C) exhausted their supply of nutrients
    (D) were exposed to high temperature
    (E) were exposed to increased pH

83. During which of the following time periods did the bacteria show the greatest percent increase in population?

    (A) between 0 and 5 hours
    (B) between 5 and 10 hours
    (C) between 10 and 15 hours
    (D) between 15 and 20 hours
    (E) between 20 and 25 hours

GO ON TO THE NEXT PAGE

BIOLOGY – Continued

Questions 84-87 refer to an experiment designed to assess the effect of one organism's population growth on another's. On day 1 a species of yeast was initially cultured in a growth medium under appropriate conditions in the absence of any other species. On day 3 paramecia were added to the medium and began to feed on the yeast. The researchers monitored the population size for both species and obtained findings as shown below.

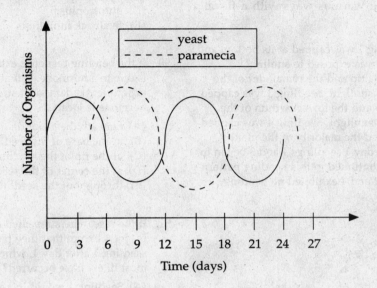

84. In terms of the experiment described above, the paramecia are best described as

    (A) parasites
    (B) saprophytes
    (C) preys
    (D) predators
    (E) producers

85. The experiment and data most strongly suggest that the periodic decline in the yeast population is due to

    (A) mutation
    (B) competition
    (C) consumption by heterotrophs
    (D) exhaustion of food supply
    (E) inability to conduct photosynthesis

86. If the paramecia destroy all the yeast population, the paramecia in this experiment will

    (A) become more abundant
    (B) become extinct
    (C) consume other organisms
    (D) increase the carrying capacity of the yeast population
    (E) reproduce at a higher rate

87. If, after adding the paramecia to the growth medium, the researchers had also added another species capable of feeding on yeast, which of the following processes would most quickly occur?

    (A) Commensalism
    (B) Saprophytism
    (C) Symbiosis
    (D) Evolution
    (E) Competition

GO ON TO THE NEXT PAGE

BIOLOGY – Continued

Questions 88–91 refer to an experiment in which investigators probed the response of 4 plant seedlings to various conditions of sunlight deprivation over a period of 4 days. Each seedling was 6 inches tall and seedlings 1 through 4 were capped in various ways with a flexible material.

As shown below, seedling 1 was capped at its bottom and the remainder of the plant was exposed to sunlight. Seedling 2 was capped at its tip and the remainder of the plant remained exposed to sunlight. Seedling 3 was capped near its tip but both the tip and the lower portion of the plant remained exposed to sunlight. Seedling 4 was capped at its tip and the cap covered the majority of the plant's body. It was found that on day 1 seedlings 1 and 3 began to bend in their growth and continued in this bending pattern through day 4. Seedlings 2 and 4 exhibited no bending.

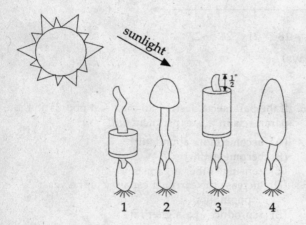

88. It is most likely that seedlings 1 and 3 bend

    (A) away from the source of sunlight
    (B) toward the source of sunlight
    (C) toward the center of the Earth
    (D) only during the night
    (E) only in conditions of high humidity

89. The bending of seedlings 1 and 3 is an example of

    (A) phototropism
    (B) geotropism
    (C) hydrotropism
    (D) autotrophism
    (E) feedback inhibition

90. If the bending response exhibited by seedlings 1 and 3 is due to a hormone activated by sunlight, the experimental data most strongly suggest that the hormone is located

    (A) beneath the ground
    (B) at the base of the seedling
    (C) at the tip of the seedling
    (D) at the center of the seedling
    (E) throughout the seedling

91. If the researchers repeated their experiment in all respects except that they placed a cap at the tip of seedling 1 after day 1, which of the following would most likely have occurred?

    (A) Seedling 1 would increase its rate of carbon dioxide consumption.
    (B) Seedling 1 would increase its rate of growth.
    (C) Seedling 1 would reverse the direction of its bending.
    (D) Seedling 1 would continue to bend throughout days 2 through 4.
    (E) Seedling 1 would exhibit no additional bending on days 2 through 4.

GO ON TO THE NEXT PAGE

BIOLOGY — Continued

Questions 92-95 refer to the data below, resulting from an experiment in which facultative anaerobic organisms were placed in a carbohydrate growth medium and subjected first to aerobic conditions and then to anaerobic conditions.

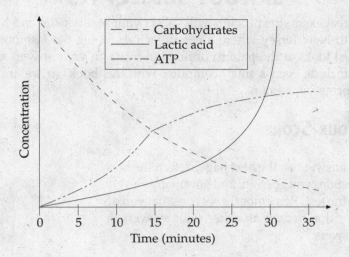

92. According to the graph shown above, which of the following is true regarding ATP at time = 15 minutes?

    (A) The concentration of ATP begins to decrease.
    (B) The rate of production of ATP begins to decrease.
    (C) The consumption of ATP begins to increase.
    (D) ATP molecules are broken down into ADP and inorganic phosphate.
    (E) ATP is phosphorylated at increased rates.

93. Which of the following is true concerning the anaerobic and aerobic respiratory processes related to the experiment?

    (A) Anaerobic respiration makes more efficient use of carbohydrate than does aerobic respiration.
    (B) Anaerobic respiration is always accompanied by photosynthesis, and aerobic respiration is not.
    (C) The rates of ATP production are equivalent for aerobic and anaerobic respiration.
    (D) The rate of ATP production is greater for aerobic respiration than for anaerobic respiration.
    (E) The rate of ATP production is greater for anaerobic respiration than for aerobic respiration.

94. Which of the following is a substance that may result from the process of anaerobic respiration?

    (A) Glycogen
    (B) Glucose
    (C) Cytochrome
    (D) Carbon monoxide
    (E) Lactic acid

95. Based on the graph shown above, conditions are altered from aerobic to anaerobic after the expiration of

    (A) 5 minutes
    (B) 15 minutes
    (C) 20 minutes
    (D) 25 minutes
    (E) 35 minutes

STOP

IF YOU FINISH BEFORE TIME IS CALLED, YOU MAY CHECK YOUR WORK ON THIS TEST ONLY.
DO NOT TURN TO ANY OTHER TEST IN THIS BOOK.

HOW TO SCORE THE PRINCETON REVIEW BIOLOGY SUBJECT TEST

When you take the real exam, the proctors will collect your text booklet and bubble sheet and send your answer sheet to New Jersey where a computer (yes, a big old-fashioned one that has been around since the '60s) looks at the pattern of filled-in ovals on your answer sheet and gives you a score. We couldn't include even a small computer with this book, so we are providing this more primitive way of scoring your exam.

Determining Your Score

STEP 1 Using the answers on the next page, determine how many questions you got right and how many you got wrong on the test. Remember, questions that you do not answer don't count as either right answers or wrong answers.

STEP 2 List the number of right answers here.   (A) _____

STEP 3 List the number of wrong answers here. Now divide that number by 4. (Use a calculator if you're feeling particularly lazy.)   (B) _____ ÷ 4 _____

STEP 4 Subtract the number of wrong answers divided by 4 from the number of correct answers. Round this score to the nearest whole number. This is your raw score.   (C) (A) _____ – (B) _____ = _____

STEP 5 To determine your real score, take the number from Step 4 above and look it up in the left column of the Score Conversion Table on page 188; the corresponding score on the right is your score on the exam.

ANSWERS TO THE PRINCETON REVIEW BIOLOGY SUBJECT TEST

Question Number	Correct Answer	Right	Wrong
1.	E		
2.	C		
3.	A		
4.	D		
5.	E		
6.	A		
7.	E		
8.	D		
9.	A		
10.	B		
11.	D		
12.	C		
13.	E		
14.	C		
15.	D		
16.	B		
17.	B		
18.	C		
19.	E		
20.	E		
21.	B		
22.	E		
23.	C		
24.	C		
25.	C		
26.	B		
27.	A		
28.	C		
29.	A		
30.	B		
31.	A		
32.	A		

Question Number	Correct Answer	Right	Wrong
33.	D		
34.	B		
35.	C		
36.	B		
37.	A		
38.	D		
39.	B		
40.	E		
41.	D		
42.	D		
43.	C		
44.	A		
45.	E		
46.	B		
47.	D		
48.	A		
49.	C		
50.	A		
51.	D		
52.	A		
53.	A		
54.	B		
55.	B		
56.	C		
57.	B		
58.	A		
59.	B		
60.	C		
61.	E		
62.	A		
63.	C		
64.	A		

Question Number	Correct Answer	Right	Wrong
65.	D		
66.	B		
67.	D		
68.	C		
69.	D		
70.	A		
71.	E		
72.	A		
73.	E		
74.	A		
75.	D		
76.	C		
77.	A		
78.	E		
79.	B		
80.	D		
81.	A		
82.	C		
83.	A		
84.	D		
85.	C		
86.	B		
87.	E		
88.	B		
89.	A		
90.	C		
91.	E		
92.	B		
93.	D		
94.	E		
95.	B		

THE PRINCETON REVIEW BIOLOGY SUBJECT TEST SCORE CONVERSION TABLE

Recentered scale as of April 1995

Raw Score	Scaled Score	Raw Score	Scaled Score	Raw Score	Scaled Score
95	800	60	610	25	440
94	800	59	610	24	430
93	790	58	600	23	430
92	790	57	600	22	420
91	780	56	600	21	420
90	770	55	590	20	410
89	760	54	590	19	410
88	760	53	580	18	400
87	750	52	580	17	400
86	750	51	570	16	390
85	740	50	570	15	380
84	730	49	560	14	380
83	730	48	560	13	370
82	720	47	550	12	370
81	720	46	540	11	360
80	710	45	540	10	360
79	710	44	530	9	360
78	700	43	530	8	350
77	700	42	520	7	350
76	690	41	520	6	350
75	690	40	510	5	340
74	680	39	510	4	330
73	680	38	500	3	330
72	670	37	500	2	320
71	670	36	490	1	320
70	660	35	490	0	310
69	660	34	480	−1	310
68	650	33	480	−2	300
67	650	32	480	−3	300
66	640	31	470	−4	290
65	630	30	470	−5	290
64	630	29	460	−6	280
63	620	28	460	−7	210
62	620	27	450	−8	210
61	610	26	440	−9	200
				−10 through −25	200

7

Explanations for the SAT II: Biology Regular Subject Test

Part A

1. Which of the following accurately states the principle(s) of the cell theory?

    I. All organisms are composed of cells.
    II. All cells arise from preexisting cells.
    III. Cells are the basic unit of biological function.

    (A) I only
    (B) II only
    (C) I and II only
    (D) II and III only
    (E) I, II, and III

E is correct. The cell theory is the basis of modern biology and the three statements are its principal components. I, II, and III all apply, and that's why E is right.

2. The plasma membrane is a semipermeable organelle composed chiefly of

    (A) lipids and sugars
    (B) sugars and proteins
    (C) proteins and lipids
    (D) lipids and carbohydrates
    (E) carbohydrates and proteins

C is correct. The plasma membrane, also called the cell membrane, is made of something called a lipid protein bilayer. This means that it's made of lipids and proteins. So, whenever you see the phrase "lipid bilayer," think "plasma membrane."

3. The process of anaerobic respiration can lead to the formation of

    (A) lactic acid or alcohol
    (B) lactic acid and amino acids
    (C) carbon monoxide and water
    (D) oxygen and carbon
    (E) ammonia and alcohol

A is correct. Anaerobic respiration refers to respiration in the absence of oxygen. Pyruvic acid is produced from glucose, but it doesn't get converted to acetyl CoA in order to enter the Krebs cycle. Instead, pyruvic acid undergoes fermentation, producing lactic acid, or alcohol and CO_2. When you see the phrases "lactic acid" or "alcohol and CO_2," think "anaerobic respiration"—or "lack of oxygen."

4. Which of the following is NOT found in a molecule of DNA?

    (A) Adenine
    (B) Deoxyribose
    (C) Phosphorus
    (D) Uracil
    (E) Thymine

D is correct. When you take the Biology Subject Test, remember this key difference between RNA and DNA. RNA has uracil instead of thymine as a base.

5. According to the heterotroph hypothesis, the atmosphere of the Earth before the beginning of life contained all of the following gases EXCEPT

    (A) hydrogen
    (B) water
    (C) ammonia
    (D) methane
    (E) oxygen

E is correct. According to the heterotroph hypothesis, life began when the Earth's atmosphere contained little or no oxygen but contained instead a great deal of hydrogen, water, ammonia, and methane.

6. The process in which a dipeptide is broken into two amino acids is called

    (A) hydrolysis
    (B) dehydration synthesis
    (C) peptide bonding
    (D) hydrogen bonding
    (E) transcription

A is correct. Proteins are formed of polypeptide chains, which are chains of amino acids. A dipeptide refers to two amino acids linked together by a peptide bond. Hydrolysis, in turn, is a process whereby a peptide bond may be broken and the constituent amino acids liberated. Water is released in the process. When you see the word "hydrolysis," think "breaking of bonds" and "release of water."

7. The amoeba, paramecium, and euglena are classified in the same phylum because all

    (A) have cell walls
    (B) reproduce by spore formation
    (C) are photosynthetic
    (D) have external digestive systems
    (E) are unicellular

E is correct. These organisms are all unicellular, and that's what they have in common. None of the other characteristics described in the options A through D are common to all of the named organisms. They reproduce not by spore formation, but by mitosis.

8. Plant cells differ from animal cells in that plant cells

    (A) perform respiration
    (B) perform protein synthesis
    (C) contain a cell membrane
    (D) contain chloroplasts
    (E) contain vacuoles

D is correct. Plant cells have all fundamental components found in animal cells except that they also have (1) a cell wall and (2) chloroplasts. The chloroplasts contain chlorophyll, the pigment necessary for photosynthesis—a series of reactions that only plants, of course, can perform.

9. Which of the following accurately describes a principal difference between the human nervous and endocrine systems?

   (A) Endocrine hormones travel through the blood and nervous signals travel through neurons.
   (B) Endocrine organs secrete neurotransmitters and nerves secrete hormones.
   (C) The endocrine system relies on chemical transmission and the nervous system does not.
   (D) The endocrine system involves myelinated tissues, and the nervous system does not.
   (E) The endocrine system is involved in homeostasis and the nervous system is not.

A is correct. The endocrine system operates by secreting hormones (chemicals) into the blood stream. They travel through the blood stream and in that way reach their target organs.

The nervous system, on the other hand, sends its signals through neurons by a process of depolarization and action potential. Between neurons are spaces called "synapses." The nervous signal travels across the synapse by means of a chemical called a neurotransmitter. The most important of these neurotransmitters is acetylcholine.

10. Which of the following is involved in the transport of food within green plants?

    (A) Tracheids
    (B) Sieve-tube cells
    (C) Cambium
    (D) Stomates
    (E) Guard cells

A is correct. Sieve-tube cells are the food-conducting cells of phloem.

11. Which of the following is true concerning producers?

    (A) They are heterotrophic.
    (B) They represent the smallest single component of the Earth's biomass.
    (C) They derive their energy from organic compounds.
    (D) They derive their energy directly from solar energy.
    (E) They are carnivorous.

D is correct. This question concerns food chains, which include producers, primary consumers, secondary consumers, tertiary consumers, and decomposers. A chain is said to "start" with producers. They take energy from the sun which is, after all, the source of all the energy we get here on Earth. So the most important characteristic of a producer is that it is photosynthetic, and another word for that is "autotrophic." So when you see the word "autotrophic," think "photosynthetic" and think "producers."

12. An mRNA molecule carries the codon sequence adenine-uracil-cytosine. Which of the following represents the complementary triplet on a tRNA?

    (A) Thymine-adenine-guanine
    (B) Guanine-adenine-thymine
    (C) Uracil-adenine-guanine
    (D) Uracil-uracil-guanine
    (E) Uracil-guanine-cytosine

C is correct. Arrange the base pairs of RNA (A, C, G, U) alphabetically and realize that the base pairs at the two "ends" of the list combine and the base pairs in the "middle" of the list combine. Adenine pairs with uracil, and cytosine pairs with guanine. The sequence adenine-uracil-cytosine, therefore, would "translate" into the sequence uracil-adenine-guanine.

13. The basic principles of genetics were established in the nineteenth century by

    (A) Jean Lamarck
    (B) Louis Pasteur
    (C) Charles Darwin
    (D) James Watson
    (E) Gregor Mendel

E is correct. The test writers would like you to know some of the names associated with key discoveries. When you think of genetics, think of Mendel and his garden peas.

14. The roots of plants serve which of the following functions?

    I. Anchorage
    II. Nutrient absorption
    III. Photosynthesis

    (A) I only
    (B) II only
    (C) I and II only
    (D) II and III only
    (E) I, II, and III

C is correct. Plant roots serve to anchor the plant to the soil and to absorb water and nutrients from it. Statements I and II apply, which means you eliminate B and D. Roots do not serve a function in photosynthesis. Rather, the *leaf* is responsible for *that* all-important process.

15. Among the following, the phylogenetic group whose members show the greatest degree of similarity is

    (A) phylum
    (B) class
    (C) order
    (D) species
    (E) family

D is correct. Phylogenetic classification may be remembered with the familiar device, "King Philip Came Over From Germany, So?" This is to remind us of the words: kingdom, phylum, class, order, family, genus, species. As we proceed down this list from kingdom to species, members have more and more in common. Species is the last word on the list and the members of a species have more in common than do the members of any other phylogenetic classification.

16. A principle difference between heterotrophs and autotrophs is that
    (A) heterotrophs produce their own nutrients and autotrophs do not
    (B) heterotrophs rely on other organisms for nutrition and autotrophs do not
    (C) heterotrophs feed only on dead organisms and autotrophs feed on living organisms
    (D) heterotrophs perform photosynthesis and autotrophs do not
    (E) heterotrophs require oxygen and autotrophs do not

B is correct. This is only a matter of remembering words. Heterotroph *means* "relies on other organisms for its nutrition." Autotroph *means* "can manufacture its own nutrients—automatically." Generally, when we're talking about autotrophs we're talking about organisms that can perform photosynthesis. Learn to associate "autotroph" with "photosynthesis." And when you're thinking about food chains, also think "producer."

17. A yeast cell may undergo reproduction asexually by a process known as
    (A) sporulation
    (B) budding
    (C) vegetative propagation
    (D) binary fission
    (E) regeneration

B is correct. Yeasts belong to the kingdom of fungi. They reproduce asexually by budding: they shed a piece of themselves, and that piece becomes a new organism. Remember to associate "yeast" with "fungus," and keep in mind that yeasts, unlike most other fungi, produce by budding.

18. Which of the following is more primitive than a bird but more advanced than an amphibian?
    (A) Amoeba
    (B) Fish
    (C) Reptile
    (D) Mammal
    (E) Slime mold

C is correct. Think **FARBM**: Fish, Amphibian, Reptile, Bird, Mammal. A reptile falls "between" a bird and an amphibian.

19. A principal difference between mitosis and meiosis is that

   (A) mitosis involves physical division and meiosis does not
   (B) mitosis involves replication of chromosomes and meiosis does not
   (C) mitosis occurs only in sex cells and meiosis occurs only in somatic cells
   (D) mitosis occurs only in males and meiosis occurs only in females
   (E) mitosis gives rise to diploid daughter cells and meiosis gives rise to haploid daughter cells

E is correct. In terms of end results, the most crucial difference between meiosis and mitosis is that meiosis produces haploid cells and mitosis produces diploid cells. The haploid cells then go on to join other haploid cells and form diploid organisms.

The meiotic process occurs in both sexes. Human spermatozoa and ova, for example, result from meiosis. They're haploid. When they get together, they form a diploid zygote which, if all goes well, produces a viable embryo and fetus.

20. Which of the following correctly characterizes the process of evolution?

   I. Populations evolve, not individuals
   II. Mutations tend to produce genetic variability
   III. It is a change in a population's allele frequencies

   (A) I only
   (B) II only
   (C) I and II only
   (D) II and III only
   (E) I, II, and III

E is correct. Evolution is based on a change in a population, not individuals. Statement I therefore is correct, which eliminates B and D. Statement II is also correct. Mutations do produce a change in population, so eliminate A. Statement III does lead to a change in a population's allele frequency. That's why E is the answer.

21. Which of the following is most properly associated with the term *pioneer organism*?

   (A) Oak tree
   (B) Lichen
   (C) Grass
   (D) Conifer
   (E) Shrub

B is correct. The question concerns ecological succession, which begins with organisms labeled as "pioneers." Among the listed organisms, lichen (a symbiotic mix of fungus and algae) frequently serve as "pioneers," growing on rock. They release substances that help to create soil and promote the growth of other organisms like mosses, which might also be termed "pioneers." Such complex and sophisticated organisms as oak trees, grasses, conifers, and shrubs come later in an ecological succession and cannot be termed pioneers.

22. A population of a single species of birds is divided by natural disaster and the two populations are subjected to different environmental conditions. After a period of 6,000,000 years it is most likely that the descendants of the two populations, if brought together, would

    (A) occupy the same niche
    (B) have identical mating seasons
    (C) be in competition for identical resources
    (D) be unable to defend themselves against predators
    (E) be unable to mate together

E is correct. The question concerns one of evolution's most fundamental tenets. If populations of the same species are separated and subjected for long periods to different conditions, they will adapt to the altered conditions. This will result in such dramatic changes in their genomes as to render it impossible, ultimately, for them to breed with descendants of those from whom they parted. They have become separate species.

23. A principal effect of insulin produced by the pancreas is to

    (A) stimulate metabolism
    (B) increase blood glucose levels
    (C) decrease blood glucose levels
    (D) regulate secondary sex characteristics
    (E) promote bodily growth

C is correct. Insulin is a hormone produced by the pancreas. It "opens the gates" of most cells to the admission of glucose and thereby reduces the level of glucose in the blood. For the purpose of the Biology Subject Test, just remember (1) the pancreas secretes insulin, and (2) insulin reduces blood glucose levels.

24. The organ that secretes the main digestive juices for complete digestion is known as the

    (A) mouth
    (B) esophagus
    (C) pancreas
    (D) small intestine
    (E) large intestine

C is correct. Most of the enzymes that are produced for digestion are made in the pancreas: pancreatic lipase and pancreatic amylase, just to name a few. These enzymes are released into the small intestine, not produced there, so D is incorrect. The large intestine absorbs water. The mouth is involved in the digestion of starch. The esophagus does not secrete any enzymes.

25. All of the following hormones are involved in the menstrual cycle EXCEPT

    (A) progesterone
    (B) estrogen
    (C) testosterone
    (D) follicle-stimulating hormone (FSH)
    (E) luteinizing hormone (LH)

C is correct. The menstrual cycle is a female phenomenon, and testosterone is a male hormone. It's easy to pick it out as the one that doesn't belong; all of the other named hormones *are* directly related to the menstrual cycle.

26. The musculoskeletal system arises from which of the following embryonic germ layers?

    (A) Endoderm
    (B) Mesoderm
    (C) Ectoderm
    (D) Epidermis
    (E) Epididymus

B is correct. The three embryonic germ layers are endoderm, mesoderm, and ectoderm. Endoderm gives rise to the inner linings of the digestive and respiratory tracts. Ectoderm gives rise to the epidermis, nervous system, and eye. *Everything else*, including the musculoskeletal system, comes from the mesoderm. That's why B is correct.

27. The exchange of corresponding chromatid segments between homologous chromosomes occurs during

    (A) Prophase I
    (B) Metaphase I
    (C) Anaphase I
    (D) Prophase II
    (E) Metaphase II

A is correct. The exchange of pieces of the chromosome, *crossing over*, occurs during prophase I. This leads to genetic variability.

28. In a marathon runner, painful muscle fatigue may be caused by the products of

    (A) carbon fixation
    (B) aerobic respiration
    (C) anaerobic respiration
    (D) lipolysis
    (E) proteolysis

C is correct. If one exerts muscles to the point that circulation cannot afford them enough oxygen to maintain aerobic respiration, the muscles begin to respire anaerobically. The pain that results is caused by a build-up of lactic acid, the product of anaerobic respiration.

For the Biology Subject Test, remember to associate oxygen deprivation with (1) anaerobic respiration, (2) lactic acid, and (3) muscle pain.

29. Which of the following equations represents the process of photosynthesis?

    (A) Carbon dioxide + water → glucose + oxygen + water
    (B) Glucose + oxygen + water → carbon dioxide + water
    (C) Glucose + water → carbon dioxide + water + oxygen
    (D) Ammonia + water → glucose + oxygen + water
    (E) Carbon + water → glucose + carbon dioxide + water

A is correct. Photosynthesis is, in a sense, the opposite of respiration. When it comes to the reactions of photosynthesis, remember this above all else: *glucose and oxygen are products, not reactants.* Photosynthesis takes the energy of the sun, combines it with carbon dioxide (and water), and thus produces glucose (a carbohydrate nutrient), oxygen, and water. Organisms that conduct photosynthesis are termed autotrophs, and we need them in order to keep our atmosphere rich with oxygen.

30. In chick embryos, the extraembryonic membrane that functions in excretion is called the

    (A) allantois
    (B) chorion
    (C) amnion
    (D) yolk sac
    (E) egg shell

A is correct. The extraembryonic membrane that is involved in excretion—collecting wastes—is the allantois. Don't confuse this membrane with the other answer choices. The yolk sac provides food for the embryo. The amnion is the membrane that buffers the embryo. The chorion surrounds all the other membranes. The egg shell is the hard covering.

31. In humans, gas exchange occurs principally within

    (A) the alveoli
    (B) the trachea
    (C) the nephron
    (D) the lymph glands
    (E) the nostrils

A is correct. In humans, gas exchange occurs between inspired air and the gas carried to the lung's alveoli via the pulmonary capillaries. The trachea leads to the lungs which, in turn, lead to the alveoli. However, it is at the alveoli that gas exchange occurs.

32. Which of the following exhibit an exoskeleton?

    (A) Insects
    (B) Birds
    (C) Fish
    (D) Worms
    (E) Mammals

A is correct. Insects are arthropods. Among the most significant differences between arthropods and vertebrates is the fact that arthropods have jointed appendages and an exoskeleton—a skeleton that lies outside the body instead of inside. For the Biology Subject Test, know that vertebrates have endoskeletons (skeletons inside the body) and that arthropods have exoskeletons. Examples of arthropods: grasshoppers and lobsters.

33. For a guinea pig, black coat (B) is dominant over white coat (b). If two guinea pigs mate and produce 75% offspring with black coats and 25% offspring with white coats, then the genotypes of the parent organisms are most likely

    (A) BB × BB
    (B) Bb × BB
    (C) BB × bb
    (D) Bb × Bb
    (E) bb × Bb

D is correct. We're dealing with a typical Biology Subject Test-type Mendelian cross. In order for any of the offspring to show the recessive trait, they'd have to be homozygous for it. That means that *both* parents would have to be carrying the recessive gene. Knowing that, you can eliminate A, B, and C. We're left with D and E.

Now, with the usual Punnett square analysis, figure out what progeny would result from the crosses shown in choices D and E. The cross in E would result in Bb, bb, Bb, and bb; that's 50% dominant trait (black coats) and 50% recessive trait (white coats). The cross in D results in BB, Bb, bB, and bb; that's 75% dominant trait (black coats) and 25% recessive trait (white coats). That's just what we're looking for, so D is right.

34. All of the following represent correctly paired organic compounds and the products of their decomposition EXCEPT

    (A) protein → amino acids
    (B) glucose → lactic acid
    (C) lipids → fatty acids
    (D) cellulose → carbohydrates
    (E) starch → simple sugars

B is correct. Glucose does not decompose into lactic acid. (Anaerobic respiration of glucose may produce lactic acid as a byproduct, but glucose isn't composed of lactic acid and hence cannot decompose to produce it.) Glucose is a carbohydrate. All of the other paired substances do represent organic compounds and the smaller entities from which they are made.

35. In humans, which of the following correctly represents the sequence of embryological development?

    (A) fertilization → zygote → gametes → embryo → cleavage
    (B) fertilization → gametes → cleavage → zygote → embryo
    (C) gametes → fertilization → zygote → cleavage → embryo
    (D) cleavage → gametes → embryo → fertilization → zygote
    (E) zygote → fertilization → embryo → gametes → cleavage

C is correct. In order to promote a developing embryo, we start with gametes (sperm and an ovum). The sperm fertilizes the ovum and a zygote results. The zygote undergoes cleavage and implants in the uterus. There it grows and develops into an embryo. So the order of events is:

gametes → fertilization → zygote → cleavage → embryo

36. Which of the following represents the approximate ratio of carbon to hydrogen in a molecule of sucrose?

    (A) 1 : 1
    (B) 1 : 2
    (C) 1 : 3
    (D) 2 : 1
    (E) 3 : 1

B is correct. Sucrose is a disaccharide (the combination of a molecule of fructose and a molecule of glucose). It's formula is $C_{12}H_{22}O_{11}$. That means the ratio of carbon to hydrogen is approximately 1:2.

37. Parathormone, which is produced by the parathyroids, regulates the metabolism of

    (A) calcium
    (B) glucose
    (C) potassium
    (D) iodine
    (E) zinc

A is correct. When you see "parathyroid," think "calcium" and don't confuse it with the thyroid. When you see "thyroid" (without the "para"), think "iodine." The thyroid gland requires iodine for the manufacture of its hormone, and the Biology Subject Test writers seem to think that's important. Parathormone, which comes from the parathyroid gland, regulates calcium metabolism. It increases the calcium level in the blood stream by "instructing" (a) the intestines to absorb more calcium, (b) the bones to release some calcium, and (c) the kidneys to refrain from excreting calcium.

38. All of the following compounds are organic EXCEPT
    (A) glycogen
    (B) fats
    (C) maltose
    (D) water
    (E) starch

D is correct. All of the other options are organic compounds; they contain carbon. Water (H₂O) does not contain carbons.

39. Which of the following graphs best describes the relationship between reaction rate as a function of the amount of substrate if the amount of enzyme is limited?

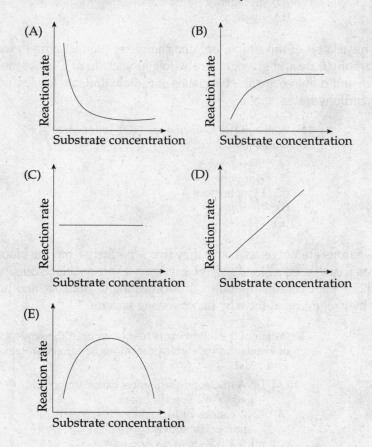

B is correct. Before you roam around the answer choices, make sure you understand the question. Someone's running a chemical reaction. She's going to start with substrate (reactant) and enzyme. Then she's going to keep adding substrate without adding any additional enzyme.

Now, with that in mind, what do you expect will happen? At first the enzyme is not saturated. So addition of substrate will speed the reaction along. The substrate will "use" the enzymes that are standing idle. But once the experimenter has added sufficient substrate to make use of all the enzymes—to saturate the enzymes—additional substrate won't speed the reaction because all the enzymes will be busy and won't be able to match up with new substrate.

All of that means you're looking for a graph that shows (1) an initial increase in reaction rate with initial addition of substrate and then (2) a "flat-line" rate even though the experimenter continues to add substrate. That's exactly what graph B shows. The others show nothing of the kind.

Once again, if you take the trouble to understand what you're being asked, you won't find yourself wandering aimlessly among the answer choices deciding to choose one that somehow "looks good."

40. All of the following compounds are carbohydrates EXCEPT

    (A) cellulose
    (B) starch
    (C) glucose
    (D) glycogen
    (E) lipid

E is correct. This is a bit of the simple organic chemistry the test writers want you to know about. Glucose is a carbohydrate and glycogen is a whole long chain of glucose molecules. That eliminates C and D. Starch and cellulose are also big chains of carbohydrate molecules, which eliminates A and B. Lipids are *not* carbohydrates.

41. Which of the following are NOT found circulating in human blood?

    (A) Erythrocytes
    (B) Hormones
    (C) Lymphocytes
    (D) Neurons
    (E) Platelets

D is correct. Neurons are nerve cells, and they live separately from the blood (except that they, like all living human tissue, need a blood *supply*). Erythrocytes are red blood cells, lymphocytes are one form of white blood cell, platelets run around the blood helping to clot cuts, and hormones run around the blood because they're unloaded there by the endocrine system.

42. Which of the following is first to occur in the sequence of events through which chromosomes direct protein synthesis?

    (A) DNA nucleotide sequences cause amino acids to gather in a specific order.
    (B) RNA codons cause amino acids to gather in appropriate order.
    (C) mRNA travels to ribosomes.
    (D) DNA generates mRNA.
    (E) tRNA anticodon attaches to mRNA codon.

D is correct. When we speak of protein synthesis we think of these events

- transcription, in which DNA generates mRNA
- translation, in which mRNA causes tRNA codons to line up in their appropriate order
- polypeptide formation/protein synthesis, in which the amino acids that are attached to tRNA bond together.

So what's the first step in the sequence? Transcription (DNA to mRNA), and that's just what choice D tells us.

The last two parts happen on ribosomes, which are located on rough endoplasmic reticulum. So when you think "protein synthesis," don't forget to think "ribosomes" and "rough endoplasmic reticulum."

43. In mammalian species, epinephrine and norepinephrine are released by the
    (A) heart
    (B) muscles
    (C) adrenal medulla
    (D) thyroid
    (E) pancreas

C is correct. Epinephrine and norepinephrine (known also as adrenaline and noradrenaline) are released by the adrenal medulla, and biologists like to say they're responsible for the "fight-or-flight" response. These hormones get you ready for action.

All of this in turn is connected to the autonomic nervous system and, in particular, to its <u>sympa</u>thetic as opposed to its <u>para</u>sympathetic component. If you don't feel sure about what all those words really mean, that doesn't matter one bit. Just remember: sympathetic system goes with epinephrine and norepinephrine, and those hormones prepare your body for action.

44. Which of the following correctly orders the events associated with the discharge of a human neuron?
    (A) polarization → depolarization → action potential → repolarization
    (B) repolarization → action potential → polarization → depolarization
    (C) action potential → polarization → repolarization → depolarization
    (D) depolarization → repolarization → action potential → polarization
    (E) depolarization → polarization → action potential → repolarization

A is correct. The unfired neuron is in a state of *polarization*; the inside is negatively charged relative to the outside. When the neuron is adequately stimulated, sodium ions rush in and the neuron is depolarized. Somehow or other, when one area of the neuron is depolarized, the whole process spreads right along the neuron like an epidemic. That's called an action potential. As it spreads, however, each section that's just been depolarized becomes repolarized. And for the Biology Subject Test, that's all you have to know.

So, the sequence of events is

polarization → depolarization → action potential → repolarization

45. Which of the following structures are found among bacteria?

   I. Plasma membrane
   II. Cell wall
   III. DNA

   (A) I only
   (B) II only
   (C) I and II only
   (D) II and III only
   (E) I, II, and III

E is correct. Bacteria are prokaryotes which have a plasma membrane, a cell wall and DNA (although it isn't enclosed in a nuclear membrane). All three statements are therefore correct.

Questions 46-50 refer to the enzymatic breakdown of maltose as shown schematically below.

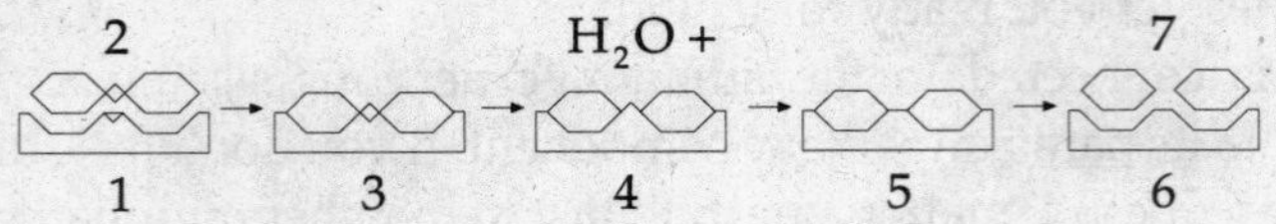

46. Which structure represents a substrate not yet attached to its enzyme's active site?

   (A) 1
   (B) 2
   (C) 4
   (D) 5
   (E) 7

B is correct. The sequence of illustrations shows (A) a substrate unbound to its enzyme, (B) a substrate-enzyme complex, (C) reaction progress (with the addition of water), and (D) separation of the products from the enzyme, leaving the enzyme intact, as it should be. Structure 2 is the substrate not yet attached to its enzyme.

47. Which structure is identical to structure 1?

   (A) 3
   (B) 4
   (C) 5
   (D) 6
   (E) 7

D is correct. Structure 6 is the enzyme molecule detached from the products of reaction. One of the most important points about enzymes is this: they are not consumed during the course of the reaction, and when the reaction is all over they're just the same as they were before it began. So, structure 6 is identical to structure 1.

48. If structure 1 were replaced with another substance, the reaction would not occur due to the phenomenon of

   (A) enzyme specificity
   (B) denaturation
   (C) dehydration synthesis
   (D) fermentation
   (E) peptide bond formation

A is correct. When we talk about enzymes, we have to remember that enzymes are picky about which reactions they'll catalyze. They only "like" certain substrates. All of that is called enzyme specificity. If we replaced structure 1 (the enzyme) with some other substance, the reaction would not occur. Enzymes are highly particular as to which substrates they'll "work with."

49. According to the schematic diagram presented above, the enzymatic breakdown of maltose requires the addition of

   (A) oxygen
   (B) hydrogen
   (C) water
   (D) carbon dioxide
   (E) adenosine triphosphate

C is correct. The illustration clearly demonstrates that water must be added to the brew in order for this reaction to proceed. You should remember that this is frequently true of breakdown/decomposition biological reactions. When things are broken apart, like proteins or fats, water is often required. Remember hydrolysis: "hydro" for water and "lysis" for splitting things apart.

50. Among the following, which term best describes the enzyme that catalyzes the breakdown of maltose?

   (A) Polypeptide chain
   (B) Amino acid
   (C) Carbohydrate
   (D) Starch
   (E) Nucleic acid

A is correct. All enzymes are proteins and all proteins are chains of amino acids. Another name for an amino acid chain is a "polypeptide chain," so that's why A is right. B is wrong because an enzyme is not composed of a *single* amino acid, it's composed of a long *chain* of amino acids that make a protein.

Questions 51-54 refer to the schematic diagram of a human being, with endocrine organs shown in place.

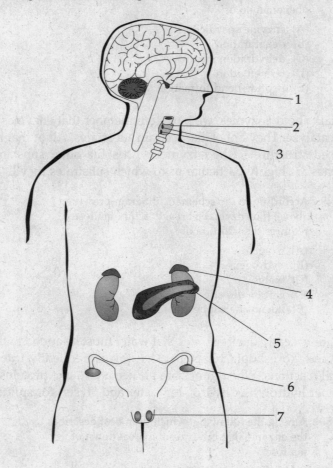

Male and Female Organs are shown

51. Which secretes a hormone that reduces blood levels of glucose?
    (A) 1
    (B) 2
    (C) 4
    (D) 5
    (E) 7

D is correct. Organ 5 is the pancreas. Its job? To secrete insulin, which (A) takes glucose out of the blood and (B) sends it into the cells. When you take the Biology Subject Test and see "insulin," remember "pancreas/lowers blood glucose level."

52. Which organ secretes thyroxine?

(A) 2
(B) 3
(C) 4
(D) 5
(E) 6

A is correct. Organ 2 is the thyroid. What does it do? It secretes thyroxine which regulates metabolism. Don't confuse the thyroid with the parathyroid, organ 3.

53. Which gland has an anterior portion that secretes adrenocorticotropic hormone (ACTH) and a posterior portion that secretes antidiuretic hormone (ADH)?

(A) 1
(B) 2
(C) 4
(D) 5
(E) 7

A is correct. Structure 1 is the pituitary gland, which has an anterior and posterior portion. The mere reference to "anterior" and "posterior" should tip you off that we're talking about the pituitary. The pituitary is located in the brain. Its anterior portion secretes a whole bunch of hormones including ACTH (which stimulates the adrenal cortex to produce *its* hormones). Its posterior portion secretes ADH (antidiuretic hormone) and oxytocin. ADH causes the kidneys to retain water (concentrate urine), and oxytocin causes the uterus to contract during childbirth.

54. Which gland has an outer cortex that secretes glucocorticoids and mineralocorticoids and an inner medulla that secretes epinephrine and norepinephrine?

(A) 2
(B) 4
(C) 5
(D) 6
(E) 7

B is correct. As soon as we see the words "cortex" and "medulla," we know we're talking about the adrenal glands, which sit on top of (but are not a part of) the kidneys. The outer portion (cortex) secretes mineralocorticoids (which cause the kidneys to retain water) and glucocorticoids (which causes the liver to produce and secrete glucose). The inner portion (medulla) secretes epinephrine and norepinephrine. These two hormones serve the sympathetic nervous system and get us ready for the famous "flight-or-fight" response (action): Heart and breathing rate goes up, and blood vessels constrict.

Part B

Questions 55-58 refer to the following diagram of a portion of the human female anatomy.

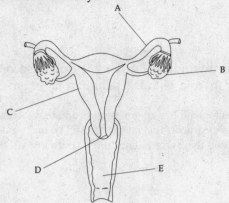

55. Location at which maturation of gametes occurs

B is correct. Structure B is the ovary and it produces ova—egg cells. Ova are the female gametes. (Sperm cells are the male gametes.)

56. Structure in which embryo is implanted

C is correct. Structure C is the uterus. When a sperm (a gamete) fertilizes an ovum (a gamete), which usually happens in the fallopian tube (structure A), the resulting zygote implants in the uterus and develops from an embryo to a fetus.

57. Structure in which estrogen is produced

B is correct. Structure B designates the ovary (there's one on each side), and the ovary produces not only ova (egg cells) but also a couple of hormones, including estrogen.

58. Site of fertilization of ovum

A is correct. Structure A designates the fallopian tube, which is also known as the oviduct. There's one on each side, and it's the tube through which the ovum travels once it's released from the ovary (structure B). It's actually in the oviduct that fertilization usually takes place.

Questions 59-62

(A) Right atrium
(B) Aorta
(C) Pulmonary artery
(D) Right ventricle
(E) Left atrium

59. Artery through which oxygenated blood leaves heart to be carried throughout body

B is correct. The aorta is the body's largest artery. It takes blood from the left ventricle and then, by dividing over and over again, it delivers oxygen-rich blood throughout the body. The blood in the aorta is oxygen-rich as opposed to the blood in the pulmonary artery, which carries blood from the right atrium to the lungs. There it can pick up a fresh supply of oxygen.

60. Artery that carries deoxygenated blood from the heart

C is correct. The pulmonary artery leaves the heart's right ventricle carrying oxygen-poor blood to the lungs. In the lungs and—in particular—at the alveoli, the blood picks up a fresh supply of oxygen. Meanwhile it dumps carbon dioxide into the inspired air. So remember, pulmonary artery-poor (in oxygen) and aorta-affluent when it comes to oxygen.

61. Chamber in heart that receives blood directly from pulmonary circulation

E is correct. The pulmonary circulation means all of the branches of the pulmonary artery that break up into tiny capillaries and pick up oxygen from the alveoli (meanwhile dumping carbon dioxide *into* the alveoli). The capillaries all come together to form one big vein, the pulmonary vein, which carries oxygen-rich blood to the heart's left atrium. (From there the oxygen-rich blood is moved straight to the heart's left ventricle and pumped out the aorta.)

62. Chamber in heart that receives deoxygenated blood returning from locations throughout the body

A is correct. After blood has coursed the entire body, delivering oxygen and picking up carbon dioxide, it enters the heart's right atrium via the anterior and posterior vena cavae. (From there it's pumped into the right ventricle and then out the pulmonary artery so it can reach the lungs and pick up a fresh supply of oxygen.)

Questions 63-65

(A) Tundra
(B) Tropical forest
(C) Taiga
(D) Desert
(E) Temperate deciduous forest

63. Fauna includes moose and black bear

C is correct. The taiga is a class of biome—biological habitat—located in North America, Europe, and Asia. It has short summers and extremely cold, long winters. Flora (plant life) consists of conifers, or evergreens.

64. Flora consists mainly of lichen, mosses, and grasses

A is correct. The tundra, a class of biome—biological habitat—is located in the northern part of North America, Europe, and Asia. Soil is frozen year-round. Plant life (flora) is limited primarily to lichen, mosses, wildflowers, and grasses. For the Biology Subject Test, remember that trees generally do not grow in the tundra. (They do grow, however, in the taiga.)

65. Contains drought-resistant shrubs and succulent plants

D is correct. Common sense tells you that the plant life in the desert better be drought-resistant and it had better store water, which it does. The plants in the desert are called "succulent," which *means* that they store water.

Questions 66-68:

(A) Ribosomes
(B) Mitochondria
(C) Nucleus
(D) Lysosomes
(E) Cell membrane

66. Contains the enzymes of the Krebs cycle

B is correct. The mitochondria are the sites of aerobic cellular respiration. They're the cell's "powerhouse." It's there that the cell performs the aerobic phases of respiration: the Krebs cycle, oxidative phosphorylation, and the electron transport chain, all to produce ATP, which is the cell's so-called energy currency. When you take the Biology Subject Test and see "mitochondria," think "cellular respiration," "Krebs cycle, the electron transport chain, oxidative phosphorylation," and, of course—"ATP production." The enzymes for the Krebs cycle would also be found in the mitochondria.

67. Contains hydrolytic enzymes that participate in intracellular digestion

D is correct. Lysosomes are the cell's garbage disposals and compactors. The prefix lys tells you that they take things apart and destroy them. Lysosomes have hydrolytic enzymes which destroy worn out organelles and materials.

68. Site of DNA transcription

C is correct. If there's anything you've heard a hundred times, it's this: all of the cell's genetic information is contained in the nucleus. Specifically, it's within the chromosomes, which as you know are located in the nucleus. All cells of a single individual organism (that's *organism, not* species) have the same set of genes. So the genetic information in every cell of anyone's body is the same as that in every other cell of that same individual's body. DNA is transcribed (copied) in the nucleus.

Questions 69-71 refer to the following diagram of a flowering plant.

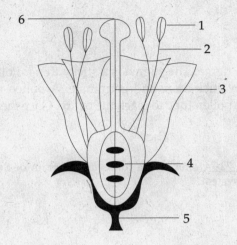

69. Which structure contains female monoploid nuclei?
    (A) 1
    (B) 2
    (C) 3
    (D) 4
    (E) 5

D is correct. Structure 4 is the flower's ovary. Inside are things called ovules. Ovules have a haploid (monoploid) number of chromosomes. So, when it comes to the Biology Subject Test and you see a flower's ovary—think "ovules" and "monoploid." The ovary and its ovules form the female part of the flower. When ovules are fertilized by pollen grains, they form a zygote. In "self-pollination" pollen from the anther (structure 1) of a flower is transferred to the stigma (structure 6) of the *same* flower. In "cross-pollination" the pollen comes from one flower and the stigma belongs to a *different* flower.

70. Pollen grains are produced by
    (A) 1
    (B) 3
    (C) 4
    (D) 5
    (E) 6

A is correct. Structure 1 depicts the anther. That's where a flower produces and stores its pollen. Pollen grains are haploid. The pollen is the male part of the plant, and it fertilizes the ovules located in the ovary. How does it get there? It falls on a stigma (structure 6), sits there a while, causing a pollen tube to develop, and then finds its way down the pollen tube to meet up with the "female" ovule. When the pollen grain first falls on the stigma, it is said to "germinate." It gets itself ready to fuse its monoploid nucleus with an ovule's monoploid nucleus.

71. Pollen grains germinate on structure
   (A) 1
   (B) 2
   (C) 4
   (D) 5
   (E) 6

E is correct. Structure 6 is the stigma. The <u>stigma</u> is a <u>sticky</u> thing. Pollen grains that come from the anther fall on the stigma. They stick there and germinate. A pollen tube develops and the pollen grains find their way down the pollen tube to reach the ovary. Once there, they fertilize the ovule and form a zygote.

<u>Questions 72-74</u> refer to the following diagram of the human digestive system.

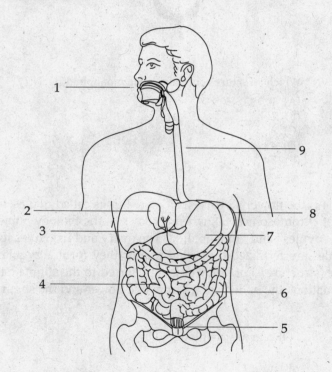

72. Bile is stored in
   (A) 2
   (B) 3
   (C) 4
   (D) 7
   (E) 8

A is correct. Bile is produced in the liver and *stored* in the gallbladder, which is located near the liver. The test writers love you to know that "the gallbladder stores bile."

73. Insulin is produced in
   (A) 1
   (B) 3
   (C) 4
   (D) 7
   (E) 8

E is correct. The pancreas, located just behind the stomach, produces insulin. What does insulin do? It reduces blood glucose levels by sending glucose into the cells. When it comes to the Biology Subject Test, remember to associate: pancreas–insulin–reduced blood glucose levels.

74. Amylase is first secreted and carbohydrate digestion begins in
   (A) 1
   (B) 2
   (C) 4
   (D) 6
   (E) 9

A is correct. The test writers like you to know that digestion begins in the mouth and, specifically, that it begins with an enzyme called amylase. Amylase digests starch, which is a carbohydrate.

Part C

Questions 75-77: The following pedigree traces the occurrence of a recessive trait in a number of families.

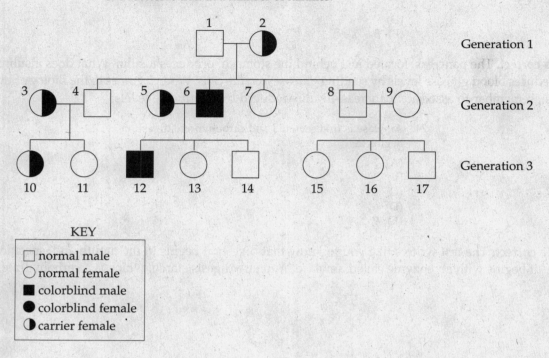

KEY
☐ normal male
○ normal female
■ colorblind male
● colorblind female
◐ carrier female

75. Among the following, what would constitute evidence that the trait is sex-linked?

(A) The trait never occurs in females.
(B) The trait never occurs in males.
(C) The trait never passes from a male parent to a female child.
(D) The trait never passes from a male parent to a male child.
(E) The trait never passes from a female parent to a male child.

D is correct. If a trait is sex-linked, it's carried on the X chromosome. A male parent may give an X chromosome, but if he does, that child is female (since the female gives only an X chromosome). If a male parent gives a Y chromosome, thus producing a male child, he cannot pass on a sex-linked trait—the trait isn't *on* the Y chromosome.

76. The pedigree shown above indicates that if individual 10 bears a child with a colorblind male, the probability that the child will be colorblind is

    (A)   0%
    (B) 25%
    (C) 50%
    (D) 75%
    (E) 100%

C is correct. Individual 10 is a female who "carries" the colorblind trait. That means that for this recessive trait, one of her alleles is normal and the other is not. Let's say capital "X" stands for the normal dominant trait and small "X_c" stands for the recessive colorblind trait. If individual 10 mates with a colorblind male, here's what we've got:

XX_c crossing with $X_c Y$.

In terms of probabilities, that gives us:

XX_c XY $X_c X_c$ $X_c Y$.

That means that among four possible offspring genotypes,
- one will be a female carrier
- one will be a normal male
- one will be a colorblind female and
- one will be a colorblind male

So, two of four possible outcomes are colorblind offspring—one male and one female. That's 50%, and that's why C is right.

77. According to the pedigree shown above, which of the following groups of individuals is heterozygous for the colorblind trait?

    (A) Individuals 2, 3, 5, and 10
    (B) Individuals 2, 5, 8, 10, and 12
    (C) Individuals 3, 5, 9, 11, and 13
    (D) Individuals 2, 3, 5, 6, and 16
    (E) Individuals 2, 3, 4, 10, and 14

A is correct. When we say "heterozygous" in this case, we're talking about a person who has one normal gene and one abnormal gene. If the normal gene is dominant, then the person's phenotype is normal. When it comes to sex-linked traits, females are the only heterozygous people there are. (Males can't be heterozygous for sex-linked traits. They can't *have* the normal gene because they only have one X chromosome, where the gene for the trait is located.)

Therefore, to answer this question, we just look for the half-filled circles that, according to the key, represent "carrier females." Among the options A through E, only choice A presents a list of half-filled circles. That's why it's correct.

Questions 78-80

An experimenter wishes to study the manner in which newly hatched birds acquire attachments to their mothers. Six newly hatched ducklings were selected. At the moment of birth, ducklings 1, 2, and 3 were removed from the birth site so that they had no exposure to their mother. Ducklings 4, 5, and 6 remained with the mother.

A balloon was then passed by ducklings 1, 2, and 3 and simultaneously each duckling was exposed to a recording of a mother duck call. It was found over subsequent hours and days that ducklings 4, 5, and 6 imitated their mother's behavior and consistently followed her about. It was also found that ducklings 1, 2, and 3 exhibited behavior similar to that of the balloon they had seen and attempted to follow all balloons to which they were subsequently exposed.

78. The process that occurred in relation to ducklings 1, 2, and 3 is termed

(A) innate behavior
(B) reflex behavior
(C) instinct
(D) conditioned response
(E) imprinting

E is correct. When an animal makes an important "decision" early in life on the basis of something it's seen, heard, touched, or smelled, then we say that it's learned by imprinting. That's exactly what's happened to ducklings 1, 2, and 3. They saw a balloon, they heard a duckling call, and they decided that the balloon was their mother. That's a pretty important decision.

79. Ducklings 4, 5, and 6 would best be described as

(A) experimental models
(B) controls
(C) dependent variables
(D) independent variables
(E) experimental sets

B is correct. The test-writers want you to know about experiments and "controls." When an experimenter decides to do something unusual to an animal or plant or anything else, and assess its effect, then he also should leave one or more experimental subjects in the "normal" circumstance. For the Biology Subject Test, think of those that are left in the normal circumstance as the "controls." Here, ducklings 4, 5, and 6 were left in a normal way with their mothers. They're the controls.

80. If, instead of exposing ducklings 1, 2, and 3 to a balloon and a recorded duck call, the experimenter had exposed them to a cat and a recorded duck call, which of the following would most likely occur in relation to ducklings 1, 2, and 3?

    (A) They would fear other ducklings.
    (B) They would lose their ability to reproduce.
    (C) They would acquire an ability to make sounds resembling those of a cat.
    (D) They would regard the cat as their mother.
    (E) They would believe themselves to be cats.

D is correct. The test-writers want to assess your ability to think logically. Use common sense. If the ducklings thought the balloon was their mother and the balloon is replaced by a cat, then probably, they'll think the cat is their mother. Before wandering around the answer choices, you should already have *known* what kind of answer you were looking for: one that says the ducklings would treat the cat as their mother. That's just what choice D says and that's why it's right.

If you did not first know what you were looking for, you might fall right into choice E, which sets a temptation trap. Don't do it. *First* know what you're looking for and *then* look at the answer choices.

Questions 81-83 refer to the data, set forth below, resulting from an experiment concerning the rate of growth in a bacterial population over a 25-hour period.

Column I	Column II
Time (hrs)	Number of bacteria
0	750
5	9,000
10	44,000
15	35,000
20	11,000
25	6,000

81. Which of the following graphs, plotting time against bacterial population, best represents the results set forth in the table above?

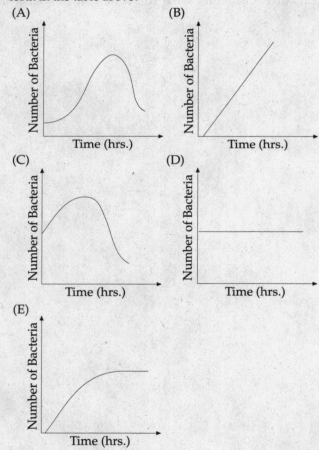

A is correct. The question tests your ability to convert data from a table to data on a graph. The question tells you that the graphs you'll be examining plot time against bacterial population. Before you even look at them and start "shopping," you should notice what the table tells you.

From 0 to 10 hours, the population increases. Then as time goes on, it decreases. So when you go to the answer choices, look for a graph that shows how population

- first increases *significantly* with time

and

- then decreases *significantly* with time.

Graph A is the only one that does that, and that's why A is right.

82. The most probable explanation for the numerical decline observed for the last three entries in column 2 is that the bacteria

   (A) exhausted their ability to manufacture respiratory enzymes
   (B) exhausted their capacity to undergo binary fission
   (C) exhausted their supply of nutrients
   (D) were exposed to high temperature
   (E) were exposed to increased pH

C is correct. The test-writers want you to be thinking about "competition for resources." There's nothing in the question or in the table that would suggest changes in temperature or pH. It's true that if the bacteria lost their respiratory enzymes or their capacity to reproduce, their numbers would certainly decline. But you're asked for the most *probable* explanation for the decreasing population. Having no other information, the most likely explanation is that they started running out of food.

Now don't let the phrase "exhausted their supply of nutrients" hide from you. It's camouflage for "ran out of food." If you go into this question knowing that the most probable explanation is a food shortage, remember to keep the blinders off your brain. There's more than one way of saying the same thing.

83. During which of the following time periods did the bacteria show the greatest percent increase in population?

   (A) Between 0 and 5 hours
   (B) Between 5 and 10 hours
   (C) Between 10 and 15 hours
   (D) Between 15 and 20 hours
   (E) Between 20 and 25 hours

A is correct. This is only an arithmetic problem, and your ability to answer it depends on your understanding of percentage.

To begin with, the answer can't be C, D, or E, because they represent time periods during which the bacterial population <u>de</u>creased. So if you're paying any attention at all, you'll narrow your options to A and B. Even a wild guess would at that point give you a 50% chance of answering correctly.

Now let's look at A and B. If you don't pay attention to the question, you might fall into choice B. It's tempting, because in *absolute* terms the largest numerical increase in population did take place between 5 and 10 hours: the increase was 35,000. But we're talking about *percent* increase. Consider choice A. To increase from 750 to 9,000 is to increase by 12 times = 1200%. Consider choice B. To increase from 9,000 to 44,000 is to increase by 4.9 times = 490%. That's why B is wrong and A is right.

Questions 84-87 refer to an experiment designed to assess the effect of one organism's population growth on another's. On day 1 a species of yeast was initially cultured in a growth medium under appropriate conditions in the absence of any other species. On day 3 paramecia were added to the medium and began to feed on the yeast. The researchers monitored the population size for both species and obtained findings as shown below.

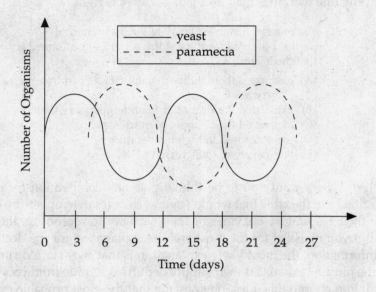

84. In terms of the experiment described above, the paramecia are best described as

    (A) parasites
    (B) saprophytes
    (C) preys
    (D) predators
    (E) producers

D is correct. In this experiment, the paramecia eat the yeast. Given the predator-prey fluctuations, the paramecium is the predator and the yeast is the prey. So C is incorrect. B is incorrect because paramecia do not feed on dead organisms. A and E are incorrect because paramecia do not live inside the yeast while feeding on it or make their own food.

85. The experiment and data most strongly suggest that the periodic decline in the yeast population is due to

    (A) mutation
    (B) competition
    (C) consumption by heterotrophs
    (D) exhaustion of food supply
    (E) inability to conduct photosynthesis

C is correct. The question tests your ability to read a graph and think logically. On day 3 the paramecia are added. They start eating the yeast, so the yeast population declines. Once the yeast population declines, the paramecia start running out of food, so *they* start competing with each other for what little food (yeast) remains. The paramecium population starts to decline for lack of food. Once *that* happens the remaining yeast are able to start repopulating, which is why their numbers start to increase when the paramecium population declines.

Now, this question asks us to explain the periodic decline in the *yeast* population, not the paramecium population. Make sure you know what you're being asked before you start touring the answer choices. Why do the yeast start disappearing periodically? They're being eaten by paramecia. You're looking for an answer choice that means "the yeast are being eaten by paramecia." That very thought is *camouflaged* in choice C: "consumption by heterotrophs," and that's why C is right. (If you were looking to explain the periodic decline in the paramecium population, B would be right, but that's *not* what you're supposed to be looking for.)

86. If the paramecia destroy all the yeast population, the paramecia in this experiment will

    (A) become more abundant
    (B) become extinct
    (C) consume other organisms
    (D) increase the carrying capacity of the yeast population
    (E) reproduce at a higher rate

B is correct. If the yeast population is destroyed by the paramecia, the paramecia will die off. A and E are incorrect because the paramecia could not increase or reproduce without food. D is incorrect because if the paramecia destroyed the yeast population, the yeast population would never reach its carrying capacity. C is absurd: there were no other organisms in the medium.

87. If, after adding the paramecia to the growth medium, the researchers had also added another species capable of feeding on yeast, which of the following processes would most quickly occur?

    (A) Commensalism
    (B) Saprophytism
    (C) Symbiosis
    (D) Evolution
    (E) Competition

E is correct. What's the most obvious thing that will happen when another species is added to the brew and starts making its "play" for the yeast? The paramecia and the new species will be competing for food. That's why competition is the right answer. It is true, by the way, that in response to the competition, the paramecia and the new species might evolve, but that takes longer. The question asks which process would most "quickly" occur. That's why D is wrong and E is right.

Questions 88-91 refer to an experiment in which investigators probed the response of 4 plant seedlings to various conditions of sunlight deprivation over a period of 4 days. Each seedling was 6 inches tall and seedlings 1 through 4 were capped in various ways with a flexible material.

As shown below, seedling 1 was capped at its bottom and the remainder of the plant was exposed to sunlight. Seedling 2 was capped at its tip and the remainder of the plant remained exposed to sunlight. Seedling 3 was capped near its tip but both the tip and the lower portion of the plant remained exposed to sunlight. Seedling 4 was capped at its tip and the cap covered the majority of the plant's body. It was found that on day 1 seedlings 1 and 3 began to bend in their growth and continued in this bending pattern through day 4. Seedlings 2 and 4 exhibited no bending.

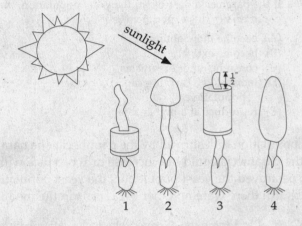

88. It is most likely that seedlings 1 and 3 bend

    (A) away from the source of sunlight
    (B) toward the source of sunlight
    (C) toward the center of the Earth
    (D) only during the night
    (E) only in conditions of high humidity

B is correct. The question concerns the effect of plant hormones called auxins. Auxins are primarily located at a plant's tip and respond to sunlight, causing the whole plant to bend toward the sun. Since the plant is bending toward its light source, this particular bending is called phototropism: "photo" for light and "trop" for growth. Know about auxins and phototropism, and you know the answer to the question. The plant bends toward the sun.

89. The bending of seedlings 1 and 3 is an example of

   (A) phototropism
   (B) geotropism
   (C) hydrotropism
   (D) autotrophism
   (E) feedback inhibition

A is correct. Auxins cause plants to bend toward the sun, and the whole process is called phototropism. Geotropism refers to a bending toward or away from the ground. Hydrotropism means growing toward a water source. Autotrophism has nothing to do with hormonal response. It refers to an organism's ability to make its own food by photosynthesis.

90. If the bending response exhibited by seedlings 1 and 3 is due to a hormone activated by sunlight, the experimental data most strongly suggest that the hormone is located

   (A) beneath the ground
   (B) at the base of the seedling
   (C) at the tip of the seedling
   (D) at the center of the seedling
   (E) throughout the seedling

C is correct. Here, the question writer assumes that you don't know that auxins are located at the seedling's tip. Rather, you are expected to do some logical thinking. The logic is pretty simple. Seedlings 2 and 4 don't bend. What else have they got in common? Their tips are covered. Seedlings 1 and 3 do bend. What else have they got in common? Their tips are not covered. It seems, then, that if a sunlight-activated hormone is responsible for the bending it must be located at the tip of the seedling. The seedlings with exposed tips do bend, and the seedlings with covered tips do not.

91. If the researchers repeated their experiment in all respects except that they placed a cap at the tip of seedling 1 after day 1, which of the following would most likely have occurred?

   (A) Seedling 1 would increase its rate of carbon dioxide consumption.
   (B) Seedling 1 would increase its rate of growth.
   (C) Seedling 1 would reverse the direction of its bending.
   (D) Seedling 1 would continue to bend throughout days 2 through 4.
   (E) Seedling 1 would exhibit no additional bending on days 2 through 4.

E is correct. Like question 90, this question calls for a little logic. We're figuring that some hormone (auxins as it happens) is located at the seedling's tip. If we capped seedling 1's tip at the end of day 1 then, probably, it would not exhibit any further bending. The sunlight would have no access to the hormone.

Questions 92-95 refer to the data, as shown below, resulting from an experiment in which facultative anaerobic organisms were placed in a carbohydrate growth medium and subjected first to aerobic conditions and then to anaerobic conditions.

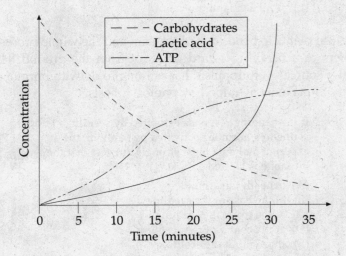

92. According to the graph shown above, which of the following is true regarding ATP at time = 15 minutes?

 (A) The concentration of ATP begins to decrease.
 (B) The rate of production of ATP begins to decrease.
 (C) The consumption of ATP begins to decrease.
 (D) ATP molecules are broken down into ADP and inorganic phosphate.
 (E) ATP is phosphorylated at increased rates.

B is correct. The ATP line has a nice upward slope until we reach time = 15 minutes. It does continue upward to some extent so that means that after 15 minutes ATP is still produced. However, the change in the slope means that ATP is not being produced at the same *rate* as it was earlier. ATP production continues, but the *rate of ATP production* declines. That's why B is right.

93. Which of the following is true concerning the anaerobic and aerobic respiratory processes related to the experiment?

 (A) Anaerobic respiration makes more efficient use of carbohydrate than does aerobic respiration.
 (B) Anaerobic respiration is always accompanied by photosynthesis, and aerobic respiration is not.
 (C) The rates of ATP production are equivalent for aerobic and anaerobic respiration.
 (D) The rate of ATP production is greater for aerobic respiration than for anaerobic respiration.
 (E) The rate of ATP production is greater for anaerobic respiration than for aerobic respiration.

D is correct. Without even referring to the graph, you know that when glucose is metabolized, aerobic respiration produces more ATP than does anaerobic respiration. Why? When there's oxygen around, the Krebs cycle, the electron transport chain, and oxidative phosphorylation can proceed. That makes for many more ATP molecules than does glycolysis alone.

We hope that the phrase "rate of ATP production" didn't hide the right answer from your view. If it did, you fell into the camouflage trap and failed to earn points for a question whose answer you definitely knew. Don't let that happen. Keep an open mind to words. Again—there's more than one way of saying the same thing.

94. Which of the following is a substance that may result from the process of anaerobic respiration?

    (A) Glycogen
    (B) Glucose
    (C) Cytochrome
    (D) Carbon monoxide
    (E) Lactic acid

E is correct. Remember: When you see "anaerobic respiration" think:

- shortage of oxygen
- lactic acid
- muscle pain

In this case, the key is "lactic acid." When glycolysis proceeds in the absence of oxygen, there's no Krebs cycle, no electron transport chain, and no oxidative phosphorylation. Rather, the pyruvic acid that glycolysis produces is converted to lactic acid in muscle cells, and ethanol (alcohol) and CO_2 in yeast.

95. Based on the graph shown above, conditions are altered from aerobic to anaerobic after the expiration of

    (A) 5 minutes
    (B) 15 minutes
    (C) 20 minutes
    (D) 25 minutes
    (E) 35 minutes

B is correct. Look at the graph. Ask yourself where things begin to change and what exactly is changing. Nothing special seems to be happening at 5, 20, 25, or 35 minutes. But at 15 minutes, two things happen. The ATP and lactic acid lines change their slopes dramatically. Why? The oxygen has been "shut off." When that happens the organism must respire <u>anaerobically</u>. That means

- it produces fewer ATP molecules per minute, because anaerobic respiration yields fewer ATP molecules from each molecule of glucose than does aerobic respiration,

    and

- it starts to produce lots of lactic acid because it can't send its pyruvic acid through the aerobic stages of respiration (the Krebs cycle, electron transport chain, and oxidative phosphorylation).

So, the graph tells us that the oxygen supply is "shut down" at 15 minutes. That's why B is right.

8
The Princeton Review SAT II: Biology E/M Subject Test

BIOLOGY E/M
SUBJECT TEST

SECTION 1

Your responses to the Biology Subject Test questions must be filled in on Section One of your answer sheet (the box on the front of the answer sheet). Marks on any other section will not be counted toward your Biology Subject Test score.

When your supervisor gives the signal, turn the page and begin the Biology Subject Test. There are 100 numbered ovals on the answer sheet and 95 questions in the Biology Subject Test. Therefore, use only ovals 1 to 95 for recording your answers.

BIOLOGY E/M TEST

Core Test

Directions: Each set of lettered choices below refers to the numbered statements immediately following it. Select the one lettered choice that best answers each question or best fits each statement, and then fill in the corresponding oval on the answer sheet. A choice may be used once, more than once, or not at all in each set.

Questions 1-3

    (A) mitochondria
    (B) cytoplasm
    (C) pyruvate
    (D) lactic acid
    (E) glucose

1. Location of cellular respiration in prokaryotes.

2. End product of anaerobic metabolism in muscle cells.

3. Location of glycolysis in eukaryotes.

Questions 4-6

    (A) anaphase II
    (B) metaphase I
    (C) prophase II
    (D) metaphase II
    (E) prophase I

4. Stage of meiosis during which recombination of genetic material occurs.

5. Stage of meiosis during which pairs of homologous chromosomes align at the center of the cell.

6. Stage of meiosis during which sister chromatids are separated.

Questions 7-9

    (A) reasoning/insight
    (B) imprinting
    (C) classical conditioning
    (D) habituation
    (E) instinct

7. A simple kind of learning involving loss of sensitivity to unimportant stimuli.

8. Geese recognize a ticking clock as "mother" if exposed to it during a critical period shortly after hatching.

9. Fish are given food at the same time as a tap on their glass bowl, and soon learn to approach when a tap sounds even in the absence of food.

Questions 10-12

    (A) small intestine
    (B) large intestine
    (C) stomach
    (D) esophagus
    (E) mouth

10. Structure where most digestion and absorption of nutrients occurs.

11. Structure where starch digestion first takes place.

12. Structure with the lowest pH.

GO ON TO THE NEXT PAGE

BIOLOGY E/M – Continued

Directions: Each of the questions or incomplete statements below is followed by five suggested answers or completions. Some questions pertain to a set that refers to a laboratory or experimental situation. For each question, select the one choice that is the best answer to the question and then fill in the corresponding oval on the answer sheet.

13. Homologous structures, which have similar underlying structures but may have different functions, are formed by

    (A) divergent evolution
    (B) speciation
    (C) segregation
    (D) convergent evolution
    (E) stabilizing selection

14. Hemoglobin is a protein in red blood cells that binds and carries oxygen and some carbon dioxide. Its affinity for oxygen changes as blood travels from the lungs to the body tissues and back to the lungs again. One could expect hemoglobin to have

    (A) a high carbon dioxide affinity in the lungs and a low oxygen affinity in the tissues
    (B) a low carbon dioxide affinity in the lungs and a high oxygen affinity in the tissues
    (C) a high oxygen affinity in the lungs and a low oxygen affinity in the tissues
    (D) a low oxygen affinity in the lungs and a high oxygen affinity in the tissues
    (E) a high oxygen affinity in the lungs and a high carbon dioxide affinity in the lungs

15. Which of the following RNA sequences would be transcribed from the DNA sequence ATGCCTAGGAC?

    (A) TACGGATCCTG
    (B) UAGCGAUCCUG
    (C) AUGCCUAGGAC
    (D) UACGGAUCCUG
    (E) GCAUUCGAAGU

16. Arthropods can be characterized by all of the following EXCEPT

    (A) a hard exoskeleton
    (B) a water vascular system
    (C) jointed appendages
    (D) molting
    (E) segmented body

17. Which of the following are functions of the kidney?

    I. filtration of blood to remove wastes
    II. blood pressure regulation
    III. pH regulation

    (A) I only
    (B) I and II only
    (C) I and III only
    (D) II and III only
    (E) I, II, and III

18. In chickens, the allele for long tail feathers (T) is dominant over the allele for short tail feathers (t). If a pure-breeding long-tailed chicken (TT) mates with a pure-breeding short-tailed chicken (tt), what percentage of their offspring (if mated with the correct genotype) could give rise to chickens with short tails?

    (A) 25%
    (B) 50%
    (C) 75%
    (D) 100%
    (E) unable to determine from the information given

19. All of the following could be considered density-dependent factors affecting population growth EXCEPT

    (A) limited nutrients
    (B) climate temperature
    (C) build-up of toxins
    (D) predation
    (E) limited water

20. The best definition of a species is

    (A) a group of organisms that occupy the same niche
    (B) a population that works together to defend itself from predators
    (C) a group of organisms that can mate with each other
    (D) a population that preys on other populations
    (E) a population where all members benefits from the association in some way

GO ON TO THE NEXT PAGE

21. Which of the following contains blood poor in oxygen?

    I. right ventricle
    II. pulmonary vein
    III. pulmonary artery
    (A) I only
    (B) II only
    (C) III only
    (D) I and II only
    (E) I and III only

22. An organism appears to be a segmented worm. Upon observation it is determined that the organism has a closed circulation, a mouth and an anus, and does NOT have an exoskeleton. The organism most likely belongs to the phylum

    (A) mollusca
    (B) annelida
    (C) echinodermata
    (D) arthropoda
    (E) chordata

23. Which of the following substances are produced by the light reactions of photosynthesis?

    (A) ATP and NADPH
    (B) ATP and glucose
    (C) NADH and glucose
    (D) ATP and NADH
    (E) NADPH and glucose

24. Consider the following graph of substrate concentration vs. product formation. Assume enzyme concentration to be constant. Why does the graph level off at high substrate concentrations?

    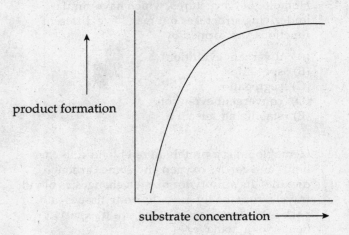

    (A) all the enzyme is used up and product formation cannot occur without enzyme
    (B) there is no more substrate to be converted into product
    (C) substrate concentration exceeds enzyme concentration and all active sites are saturated
    (D) the reaction has run to completion
    (E) an inhibitor has been added and it has slowed down the rate of product formation

25. A bird that feeds on both insects and berries would be classified as a

    I. primary consumer
    II. secondary consumer
    III. tertiary consumer
    (A) I only
    (B) II only
    (C) III only
    (D) I and II only
    (E) II and III only

BIOLOGY E/M – Continued

26. Which of the following chemical formulas could represent a carbohydrate?

    (A) $C_6H_6O_6$
    (B) $C_3H_6O_3$
    (C) $C_6H_{12}O_3$
    (D) $C_5H_{10}O_{10}$
    (E) CH_2O_4

27. A population of birds lives in an area with many insects upon which they feed. The insects live inside trees, burrowing into the bark. Over many hundreds of years, the average beak size in the bird population has increased. This is due to

    (A) increased fitness of the birds, leading to speciation
    (B) decreased fitness of the insects, allowing the birds to catch them more easily
    (C) increased fitness of large-beaked birds, leading to natural selection
    (D) decreased fitness of small-beaked birds, leading to speciation
    (E) random mutation and genetic recombination

28. The location on an enzyme where substrate binds is called the

    (A) binding site
    (B) reaction center
    (C) allosteric site
    (D) lock-and-key model
    (E) active site

29. Human cells maintain concentration gradients across their plasma membranes, such that there is a high sodium concentration outside the cell and a high potassium concentration inside the cell. Within the cell membrane are potassium "leak" channels; channels that are open all the time and through which potassium constantly exits the cell. This manner of transport across the plasma membrane is known as

    (A) simple diffusion
    (B) exocytosis
    (C) active transport
    (D) facilitated diffusion
    (E) secretion

30. The role of decomposers in the nitrogen cycle is to

    (A) fix atmospheric nitrogen into ammonia
    (B) incorporate nitrogen into amino acids and organic compounds
    (C) convert ammonia to nitrate, which can then be absorbed by plants
    (D) denitrify ammonia, thus returning nitrogen to the atmosphere
    (E) release ammonia from organic compounds, thus returning it to the soil

31. All of the following are true about the endocrine system EXCEPT

    (A) it relies on chemical messengers that travel through the bloodstream
    (B) it is a control system that has extremely rapid effects on the body
    (C) the hormones affect only certain "target" organs
    (D) it is involved in maintaining body homeostasis
    (E) its organs secrete hormones directly into the bloodstream, rather than through ducts

32. Two organisms live in close association with one another. One organism is helped by the association, while the other is neither helped nor harmed. Which of the following terms best describes this relationship?

    (A) mutualism
    (B) commensalism
    (C) symbiosis
    (D) parasitism
    (E) predator-prey relationship

GO ON TO THE NEXT PAGE

33. Cardiac output (the amount of blood pumped out of the heart in one minute) and blood pressure are directly proportional. Which of the following graphs best depicts the relationship between cardiac output and blood pressure?

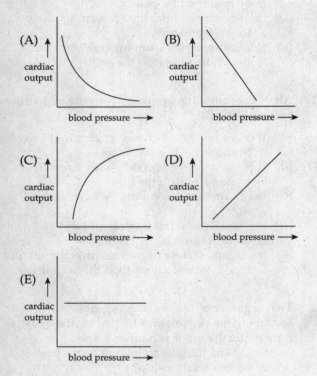

Questions 34-36 refer to the following diagram.

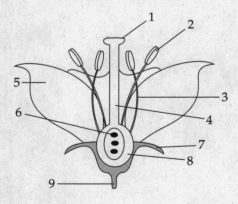

34. Location where male haploid cells are produced.
    (A) 1
    (B) 2
    (C) 3
    (D) 6
    (E) 8

35. Sticky structure where pollen grains can attach and germinate.
    (A) 1
    (B) 2
    (C) 4
    (D) 6
    (E) 8

36. Structure which, when fertilized, develops into fruit.
    (A) 1
    (B) 2
    (C) 5
    (D) 6
    (E) 8

Questions 37-38

Tropisms refer to movements made by plants, toward or away from certain stimuli. "Positive" tropisms refer specifically to movements toward a stimulus, while "negative" tropisms refer to movements made away from a stimulus.

37. A plant growing on the shady side of a building bends around the corner of the building toward the sunlight. This is an example of
    (A) negative geotropism
    (B) negative phototropism
    (C) positive phototropism
    (D) positive hydrotropism
    (E) negative hydrotropism

38. The stem and leaves of the plant grow up, away from the soil. This is an example of
    (A) negative geotropism
    (B) positive geotropism
    (C) negative phototropism
    (D) positive hydrotropism
    (E) negative hydrotropism

GO ON TO THE NEXT PAGE

BIOLOGY E/M – Continued

Questions 39-43 refer to the following diagram.

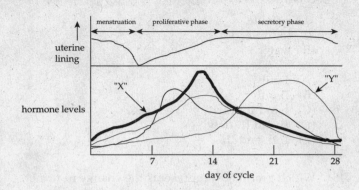

39. The hormone labeled as "X" in the diagram is often used in over-the-counter diagnostic tests to determine when ovulation has occurred. This hormone is

    (A) estrogen
    (B) progesterone
    (C) FSH
    (D) LH
    (E) testosterone

40. Based on the peak levels of hormone "X", on what day of the cycle is ovulation most likely to occur?

    (A) day 21
    (B) day 14
    (C) day 12
    (D) day 25
    (E) day 28

41. The hormone labeled as "Y" in the diagram is

    (A) progesterone, secreted by the corpus luteum after ovulation has occurred
    (B) progesterone, secreted by the ovary after ovulation has occurred
    (C) estrogen, secreted by the corpus luteum after ovulation has occurred
    (D) estrogen, secreted by the ovary after ovulation has occurred
    (E) estrogen, secreted by the follicle before ovulation occurs

42. Immediately after fertilization, the zygote begins to undergo rapid cell division. This process is known as

    (A) blastulation
    (B) gastrulation
    (C) neurulation
    (D) implantation
    (E) cleavage

43. From which of the primary germ layers does the nervous system develop?

    (A) endoderm
    (B) mesoderm
    (C) ectoderm
    (D) enteroderm
    (E) epidermis

Questions 44-46

A barren, rocky community near a lake has virtually no vegetation or animal life. After a period of approximately 75 years, the community boasts a wide variety of flora and fauna, including deciduous trees, deer, raccoon, etc.

44. The process which has taken place can best be described as

    (A) progression
    (B) succession
    (C) evolution
    (D) habitation
    (E) colonization

45. The stable community of deciduous trees and animals is known as the

    (A) final community
    (B) climax community
    (C) apex community
    (D) summit community
    (E) composite community

46. Usually the first organisms to colonize rocky areas are lichen. These are known as the

    (A) primary community
    (B) starter community
    (C) colony organisms
    (D) pioneer organisms
    (E) settler organisms

GO ON TO THE NEXT PAGE

BIOLOGY E/M – Continued

Questions 47-50 refer to the following experiment:

Diuretics are substances that help eliminate water from the body. The effects of various substances were tested on several volunteers. All volunteers had a mass of 70 kg. They drank nothing for eight hours before the test, and urinated just prior to ingesting the test substance. The three substances (water, caffeine, and salt) were tested on three separate days. The results are shown in the tables below.

Table 1

volunteer	amount caffeine ingested (in 100 ml water)	volume urine collected after 1 hour
A	50 mg	302 ml
B	100 mg	492 ml
C	150 mg	667 ml
D	200 mg	863 ml

Table 2

volunteer	amount sodium chloride ingested (in 100 ml water)	volume urine collected after 1 hour
A	0.9 mg	201 ml
B	1.8 mg	162 ml
C	2.7 mg	125 ml
D	3.6 mg	82 ml

Table 3

volunteer	volume water ingested	volume urine collected after 1 hour
A	100 mg	230 ml
B	200 mg	240 ml
C	300 mg	252 ml
D	400 mg	263 ml

47. Which of the following substances could be classified as a diuretic?

   I. caffeine
   II. sodium
   III. water

   (A) I only
   (B) II only
   (C) I and II only
   (D) II and III only
   (E) I, II, and III

48. Which graph best represents the change in urine volume when ingesting caffeine?

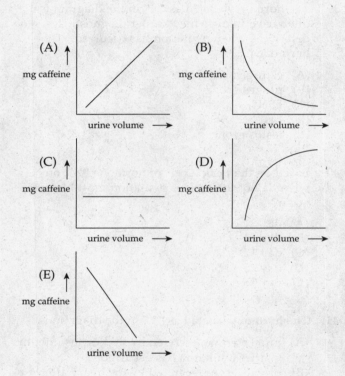

49. The purpose of ingesting the plain water was to

   (A) rehydrate the volunteers
   (B) dissolve the substances
   (C) act as a control
   (D) flush out the kidneys
   (E) act as a positive test substance

GO ON TO THE NEXT PAGE

BIOLOGY E/M – Continued

50. Based on the results in Table II, if a volunteer were to ingest 4.5 g sodium chloride dissolved in 100 ml water, what would be the approximate predicted urine volume collected after one hour?

    (A) 20 ml
    (B) 30 ml
    (C) 40 ml
    (D) 50 ml
    (E) 60 ml

Questions 51-53 refer to the following information on heredity.

Hemophilia is a disorder in which blood fails to clot. John, a male hemophiliac, marries Jane, a normal woman and together they have four children, two boys (Mark and Mike) and two girls (Molly and Mary). None of the children display the symptoms of hemophilia. Mark, Mike, Molly, and Mary all marry normal individuals and have children. None of Mark's or Mike's children, male or female, display symptoms of hemophilia, but the sons of Molly and Mary all display symptoms of hemophilia while the daughters of Molly and Mary do not.

51. Which of the following best explains the reason that Mark, Mike, Molly and Mary do not display symptoms of hemophilia, even though their father, John, is a hemophiliac?

    (A) hemophilia is an X-linked disorder, and John can only pass on his Y chromosome
    (B) hemophilia is an X-linked disorder, and even though Molly and Mary received a hemophiliac X chromosome from John, Jane gave them a normal X chromosome
    (C) hemophilia is a Y-linked disorder, and therefore cannot be displayed in females
    (D) hemophilia is a Y-linked disorder, and Mark and Mike must have received an X chromosome from John
    (E) hemophilia is an X-linked disorder, and even though Mark and Mike received a hemophiliac X chromosome from John, Jane gave them a normal X chromosome

52. If one of Mike's daughters marries a normal man, what is the probability that one of their children will display symptoms of hemophilia?

    (A) 0%
    (B) 25%
    (C) 50%
    (D) 75%
    (E) 100%

53. Which of the following individuals are heterozygous for hemophilia?

    (A) John, Mark, and Mike
    (B) Mark, Mike, Molly, and Mary
    (C) John and Jane
    (D) Molly and Mary
    (E) Mark and Mike

Questions 54-57

A volunteer was injected intravenously with several test substances to determine the effect of each substance on normal body variables. The results are shown in Table 1. Assume that enough time was allowed between injections so that the substances do not interfere with one another.

Table 1

variable	baseline values	values after injecting substand A	values after injecting substance B	values after injecting substance C	values after injecting substance D
serum Ca^{++}	2.3 mmol/L	2.3 mmol/L	3.0 mmol/L	2.3 mmol/L	2.3 mmol/L
serum Na^+	135 mmol/L	135 mmol/L	136 mmol/L	135 mmol/L	147 mmol/L
serum glucose	5.6 mmol/L	3.3 mmol/L	5.6 mmol/L	7.4 mmol/L	5.6 mmol/L

54. Based on the above information, which of the following is most likely substance B?

    (A) calcitonin
    (B) insulin
    (C) parathyroid hormone
    (D) glucagon
    (E) aldosterone

55. Based on the above information, which of the following is most likely substance A?

    (A) glucagon
    (B) aldosterone
    (C) calcitonin
    (D) parathyroid hormone
    (E) insulin

BIOLOGY E/M – Continued

56. Under what conditions might substance D be released normally?

    (A) soon after a meal
    (B) when blood pressure is low
    (C) in between meals
    (D) when there has been limited intake of dietary calcium
    (E) when dietary calcium is in excess

57. All of the following changes in variable values are significant EXCEPT

    (A) the change in serum glucose when substance A is injected
    (B) the change in serum Na⁺ when substance D is injected
    (C) the change in serum Ca⁺⁺ when substance B is injected
    (D) the change in serum glucose when substance C is injected
    (E) the change in serum Na⁺ when substance B is injected

Questions 58-60

Three different cell types were observed under the microscope. The observations are summarized in Table 1.

Table 1

Cell type	Nucleus?	Cell wall?	Chloroplasts?
A	No	Yes	No
B	Yes	Yes	No
C	Yes	Yes	Yes

The three cell types were grown in separate cultures with plenty of oxygen and nutrients available. Figure 1 shows their rates of growth. At Time 1, oxygen was no longer available to the cells.

Figure 1

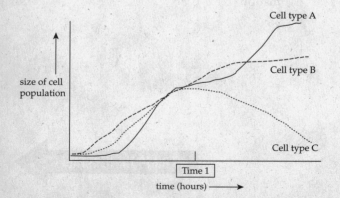

58. Based on the information in Table 1, which of the following is the most likely classification of cell Type A?

    (A) fungi
    (B) plant
    (C) bacteria
    (D) animal
    (E) protist

59. Which of the following equations is cell Type C able to run?

    I. $C_6H_{12}O_6 + 6\,O_2 \rightarrow 6\,CO_2 + 6\,H_2O + ATP$
    II. $H_2O + \text{light} \rightarrow O_2 + ATP + NADPH$
    III. $6\,CO_2 + 6\,H_2O + ATP + NADPH \rightarrow C_6H_{12}O_6$

    (A) I only
    (B) II only
    (C) I and III only
    (D) II and III only
    (E) I, II, and III

60. Consider Figure 1. Which of the following statements best describes the reason for the difference between the curves for cell Type B and cell Type C?

    (A) cell Type B is unable to survive in the presence of oxygen, while cell Type C can ferment
    (B) the products of fermentation in cell Type C are toxic to the cells and they are dying
    (C) cell Type B is an obligate aerobe while cell Type C is able to ferment
    (D) cell Type B is a facultative anaerobe, while cell Type C is an obligate aerobe
    (E) cell Type C is an obligate aerobe, while cell Type B is an obligate anaerobe

BIOLOGY E/M – Continued

BIOLOGY-E SECTION

Directions: The set of lettered choices below refers to the numbered statements immediately following it. Select the one lettered choice that best answers each question of best fits each statement, and then fill in the corresponding oval on the answer sheet. A choice may be used once, more than once, or not at all.

Questions 61-64

(A) tundra
(B) taiga
(C) chaparral
(D) deciduous forest
(E) desert

61. The driest of all terrestrial biomes, characterized by low and unpredictable precipitation.

62. Coniferous forests, characterized by long, cold winters and short, wet summers.

63. Coastal areas characterized by mild, rainy winters and long, hot, dry summers.

64. Northern areas, characterized by permafrost, extremely cold temperatures, and few trees.

Directions: Each of the questions or incomplete statements below is followed by five suggested answers or completions. Some questions pertain to a set that refers to a laboratory or experimental situation. For each question, select the one choice that is the best answer to the question and then fill in the corresponding oval on the answer sheet.

65. Plants that have true roots, stems, and leaves, as well as flowers and seeds enclosed in fruit are classified as

(A) bryophytes
(B) tracheophytes
(C) gymnosperms
(D) angiosperms
(E) endosperms

66. Which of the following indicates that animals have internal biological clocks?

(A) a mouse kept in constant darkness shows a daily rhythm of activity
(B) a rooster crows whenever the sun rises in both winter and summer
(C) an owl kept in constant light drifts away from a 24-hour cycle
(D) some species of birds can sense fluctuations in the Earth's magnetic field
(E) a squirrel whose night and day are reversed artificially soon adapts to its new schedule

67. Which of the following correctly lists the phylogenic hierarchy?

(A) kingdom, phylum, family, class, order, genus, species
(B) phylum, family, order, class, kingdom, species, genus
(C) kingdom, family, order, class, phylum, genus, species
(D) kingdom, phylum, class, order, family, genus, species
(E) family, kingdom, order, phylum, genus, class, species

68. A rattlesnake would be classified as a

(A) tertiary consumer and a heterotroph
(B) secondary consumer and an autotroph
(C) producer and an autotroph
(D) producer and a heterotroph
(E) primary consumer and a heterotroph

69. At some point in their development, chordates possess all of the following EXCEPT

(A) a dorsal hollow nerve cord
(B) a notochord
(C) gill slits
(D) postnatal tail
(E) an exoskeleton

BIOLOGY E/M – Continued

Questions 70-73

A population of birds (Population A) on a remote, isolated island is studied to determine beak length. The resulting data are plotted in Figure 1.

Figure 1

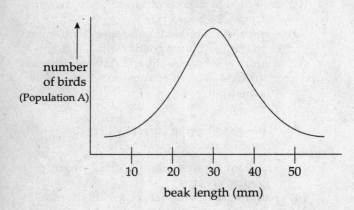

Suppose that 200 years later, the beaks of the birds on the island were again measured (Population B). The data, when plotted, yielded a graph as in Figure 2.

Figure 2

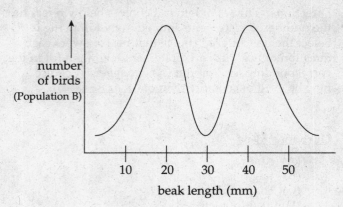

70. What is the average beak length (in cm) of the birds in Figure 1?

    (A) 30 cm
    (B) 15 cm
    (C) 5 cm
    (D) 3 cm
    (E) 1 cm

71. What is the most likely reason for the difference in distribution of beak lengths between the data plotted in Figure 1 and the data plotted in Figure 2?

    (A) all birds with beaks of 30 mm flew to a new island over the 200-year time span
    (B) birds with beaks of 30 mm were selected against
    (C) predators consumed birds with beaks of 40 mm
    (D) predators consumed birds with beaks of 20 mm
    (E) birds with beaks of 30 mm were selected for

GO ON TO THE NEXT PAGE

BIOLOGY E/M – Continued

72. Suppose that a researcher studying Population B found that birds with beaks of 20 mm were unable to mate with birds that had 40 mm beaks. These two groups of birds would now be classified as

    (A) occupying different niches
    (B) separate species
    (C) competitors
    (D) predators
    (E) separate populations

73. How might beak length in the bird population change after another 200-year time span?

    (A) the average beak length would return to 30 mm
    (B) the average beak length would shift to 40 mm
    (C) the average beak length would shift to 20 mm
    (D) the differences in beak length would be more pronounced
    (E) it is not possible to determine how beak length might change

Questions 74-78

Acid rain is formed after the burning of fossil fuels releases compounds containing nitrogen and sulfur into the atmosphere. Sunlight and rain bring about chemical reactions that convert these compounds into nitric acid and sulfur dioxide, which combine with water droplets to form acid rain. Acid rain typically has a pH of approximately 5.5.

The higher acidity of soil and water affects many living organisms adversely. As the pH of lake water falls, fish become ill and die. Table 1 shows the effects of pH on the size of adult fish.

Table 1

pH of lake	average length of fish (cm)	average mass of fish (g)
7.5	30 cm	454 g
7.0	28 cm	450 g
6.5	29 cm	453 g
6.0	25 cm	401 g
5.5	20 cm	288 g
5.0	17 cm	127 g
4.5	all fish dead	all fish dead

Mycorrhizal fungi, which form a mutualistic association with many plant roots, are particularly sensitive to the effects of acid rain. These fungi facilitate the absorption of water and nutrients by the plants; in turn, the plants provide sugars and amino acids without which the fungus could not survive.

74. The effect of acid rain on fish size is best represented by which of the following graphs?

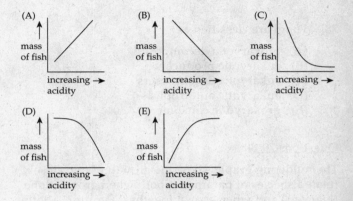

75. The relationship between mycorrhizal fungi and plants can best be described as one where

    (A) one partner benefits from the association and the other partner is harmed
    (B) one partner benefits from the association and the other partner is neither harmed nor helped
    (C) one partner preys upon the other partner
    (D) both partners benefit from the association
    (E) neither partner benefits from the association

76. If the pH of the soil were 7.0, what would be the effect on the mycorrhizal fungi and plants?

    (A) the fungi would survive but the plant would be harmed
    (B) the fungi would be harmed but the plant would survive
    (C) the fungi would be slightly harmed and the plant would be slightly harmed
    (D) neither the fungi nor the plant would survive
    (E) neither the fungi nor the plant would be harmed

GO ON TO THE NEXT PAGE

BIOLOGY E/M – Continued

77. What might be the best strategy to prevent ecological damage due to acid rain?

    (A) stock the lakes with bigger fish so that they can resist the effects of the acid better
    (B) reduce the amount of fossil fuels that are burned
    (C) supply plants with excess phosphorus and water
    (D) supply fungi with excess sugars and amino acids
    (E) only fish when it is sunny

78. Fungi are classified as

    (A) prokaryotic decomposers
    (B) eukaryotic producers
    (C) eukaryotic decomposers
    (D) eukaryotic autotrophs
    (E) prokaryotic consumers

Questions 79-80

The following graphs show the growth of two closely related species of paramecia, both when grown alone (Figure 1) and when grown together (Figure 2). Both species consume bacteria as their food source and reproduce by binary fission as often as several times a day.

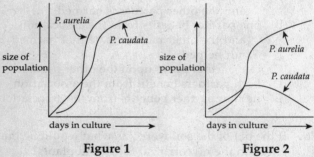

Figure 1   Figure 2

79. The data in Figure 2 indicate that

    (A) P. aurelia is preying on P. caudata
    (B) P. aurelia is a better competitor than P. caudata
    (C) P. aurelia and P. caudata are in a symbiotic relationship
    (D) P. aurelia is a parasite of P. caudata
    (E) P. aurelia grew better when combined with P. caudata that it did when grown alone

80. Paramecia are members of the kingdom

    (A) fungi
    (B) animalia
    (C) monera
    (D) protista
    (E) plantae

BIOLOGY-M SECTION

Directions: Each of the questions or incomplete statements below is followed by five suggested answers or completions. Some questions pertain to a set that refers to a laboratory or experimental situation. For each question, select the one choice that is the best answer to the question and then fill in the corresponding oval on the answer sheet.

81. All of the following is true about RNA EXCEPT

    (A) it is single stranded
    (B) its bases are adenine, thymine, guanine, and uracil
    (C) it has a sugar-phosphate backbone
    (D) its sugar is ribose
    (E) it is found in the both the nucleus and the cytoplasm of the cell

82. The function of the Golgi apparatus is to

    (A) package and store proteins for secretion
    (B) synthesize proteins
    (C) function in cellular respiration
    (D) help the cell expel waste
    (E) digest foreign substances

83. A eukaryotic cell that has a cell wall but lacks chloroplasts would be classified as a

    (A) moneran
    (B) chordate
    (C) plant
    (D) fungus
    (E) bacteria

BIOLOGY E/M – Continued

84. All of the following could give rise to new species EXCEPT

    (A) variations in antler size between male and female reindeer
    (B) an earthquake that physically separates a population of lizards into two separate groups
    (C) divergent evolution
    (D) evolution of a population of cats such that they can no longer mate with their ancestors
    (E) a massive flood that separates a population of frogs onto opposite sides of a large lake

85. The nucleic acid that is translated to make a protein is

    (A) rRNA
    (B) DNA
    (C) hnRNA
    (D) tRNA
    (E) mRNA

86. Which of the following groups have the most in common with one another?

    (A) members of the same kingdom
    (B) members of the same genus
    (C) members of the same phylum
    (D) members of the same class
    (E) members of the same family

87. Which of the following individuals is the LEAST fit in evolutionary terms?

    (A) a 45-year old male with a terminal disease who has fathered three children
    (B) a 20-year old man who has fathered one child
    (C) a 35-year old woman with four children
    (D) a healthy 4-year old child
    (E) a 25-year old woman with one child, who has had a tubal ligation to prevent future pregnancies

Questions 88-92

Most bacteria can be grown in the laboratory on agar plates containing glucose as their only carbon source. Some bacteria require additional substances, such as amino acids, to be added to the growth medium. Such bacteria are termed "auxotrophs". These bacteria are denoted by the amino acid they require followed by a "--" in superscript (e.g. arg-). Bacteria that do not require that particular amino acid can be indicated by a "+" in superscript.

Different strains of bacteria were grown on several plates containing a variety of nutrients. Figure 1 shows the colonies (numbered) which grew on each plate. The supplements in each plate are indicated.

Figure 1

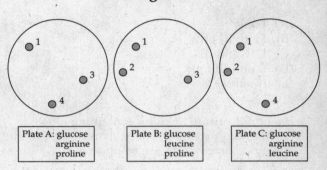

In a second experiment, Colony 1 was mixed with soft agar and spread over a plate so that an even lawn of bacteria grew. Bacterial lawns appear cloudy on agar plates. A single drop of an unknown organism was placed in the center of the bacterial lawn, and after 24 hours, a clear area known as a "plaque" appeared at that spot. The clear area continued to expand at a slow rate. Although new colonies could be grown from samples taken from the lawn, attempts to grow new colonies from samples taken from the plaque area were unsuccessful.

88. Referring to Figure 1, what is the genotype of Colony 3?

    (A) arg$^+$, leu$^+$, pro$^+$
    (B) arg$^+$, leu$^-$, pro$^+$
    (C) arg$^+$, leu$^+$, pro$^-$
    (D) arg$^-$, leu$^-$, pro$^+$
    (E) arg$^-$, leu$^-$, pro$^-$

GO ON TO THE NEXT PAGE

BIOLOGY E/M – Continued

89. Is Colony 1 an auxotroph?

    (A) Yes, it is able to grow in the presence of the three amino acids being tested.
    (B) Yes, it can only grow if glucose is present.
    (C) No, it is able to grow in the absence of glucose.
    (D) No, it is able to grow in the absence of any additional amino acids.
    (E) The data available are insufficient to determine the answer.

90. Which structures could be observed in a sample of Colony 2?

    I. nuclei
    II. ribosomes
    III. mitochondria

    (A) I only
    (B) I and III only
    (C) I, II, and III
    (D) II and III only
    (E) II only

91. If a liquid culture medium containing glucose, leucine, and proline was inoculated with Colony 4, would bacterial growth be observed?

    (A) No, Colony 4 is an arginine auxotroph (arg$^-$).
    (B) No, Colony 4 cannot grow in the presence of leucine.
    (C) Yes, Colony 4's genotype is leu$^-$, pro$^-$.
    (D) Yes, Colony 4 requires only glucose to grow.
    (E) The data available are insufficient to make a prediction.

92. What is the most likely reason for the clearing (the plaque) in the lawn of bacteria in the second experiment?

    (A) The unknown organism is bacterial Colony 2, and these bacteria are eating the bacteria from Colony 1 forming the lawn.
    (B) The unknown organism is a virus that is infecting the bacteria and causing them to lyse (killing them).
    (C) The drop placed in the center of the lawn contained a strong acid that destroyed the bacteria at that spot.
    (D) Bacteria are very delicate and the disturbance caused them to die.
    (E) The unknown organism began producing threonine, which is toxic to Colony 1.

Questions 93-96

In 1910, a small town on the East Coast of the United States relied primarily on agriculture to support its economy. In the mid-1930's, a steel mill was built, and the economy shifted from being agriculturally supported to being industrially supported. The steel mill released a lot of smog and soot into the air, which collected on the bark of trees in a wooded area near the outskirts of town. Over a period of ten years the bark gradually darkened, then maintained a constant dark color.

A variety of animals and insects lived in the wooded area. In particular, a certain species of moth served as the primary food source for a population of birds. The moths lay their eggs in the bark of the trees, and thus must spend a fair amount of time sitting on the tree trunks. Table 1 presents data on the moth population.

Table 1

Year	% white moths	% black moths
1910	95	5
1920	95	5
1930	95	5
1940	50	50
1950	20	80
1960	5	95

93. The wings of the moths and the wings of the birds are both used for flight (similar functions), however their underlying structures are very different. Moth wings and bird wings are thus classified as

    (A) homologous structures
    (B) autologous structures
    (C) divergent structures
    (D) analogous structures
    (E) emergent structures

GO ON TO THE NEXT PAGE

BIOLOGY E/M – *Continued*

94. What is the most likely explanation for the shift in the percentage of black moths in the population?

    (A) the white moths no longer blended with the color of the tree bark, and thus were selected for
    (B) the black moths blended better with the color of the tree bark, and thus were selected for
    (C) the black moths blended better with the color of the tree bark, and thus were selected against
    (D) the white moths blended better with the color of the tree bark, and thus were selected against
    (E) the black moths did not blend with the color of the tree bark, and thus were selected against

95. If a seed from one of the trees was planted in an area far from the steel mill, what color would the bark of the tree be?

    (A) black, since the parent tree had black bark
    (B) white, since the gene causing black bark was mutated due to environmental pollution
    (C) black, since the gene causing white bark was mutated due to environmental pollution
    (D) white, since the black bark was an acquired characteristic and is therefore not passed on to progeny
    (E) the color of the bark is not able to be determined

96. Birds track their prey visually, while bats rely on sonar to locate their food. If the bird population was replaced with a bat population in 1940, what would be the ratio of white moths to black moths?

    (A) 95% white, 5% black
    (B) 80% white, 20% black
    (C) 50% white, 50% black
    (D) 20% white, 80% black
    (E) 5% white, 90% black

<u>Questions 97-100</u>

Dialysis tubing is a semipermeable membrane. It allows small molecules, such as water, to pass through easily, while larger molecules, such as sucrose, are restricted. Movement of molecules across the tubing is due to concentration gradients. In an experiment designed to study osmosis, several pieces of dialysis tubing were filled with sucrose solutions of varying concentration and placed in beakers containing distilled water. The rate and direction of water movement was determined by weighing the bags before and after placing them in the distilled water. The data is recorded below.

Table 1

tube number	tube contents (beaker contents)	mass (g) 0 minutes	mass (g) 15 minutes	mass (g) 30 minutes	mass (g) 45 minutes	mass (g) 60 minutes
1	distilled water (distilled water)	22.3 g	22.4 g	22.2 g	22.3 g	22.3 g
2	10% sucrose (distilled water)	24.8 g	25.3 g	25.7 g	26.4 g	26.9 g
3	40% sucrose (distilled water)	25.1 g	26.3 g	27.5 g	28.9 g	29.6 g
4	distilled water (40% sucrose)	22.7 g	21.3 g	20.5 g	19.8 g	18.7 g

97. Why does the mass of Tube 3 increase, while the mass of Tube 4 decreases?

    (A) water is moving into Tube 3, and sucrose is moving into Tube 4
    (B) water is moving into Tube 4, and sucrose is moving into Tube 3
    (C) water is moving into Tube 3, and water is moving out of Tube 4
    (D) sucrose is moving into Tube 3, and sucrose is moving out of Tube 4
    (E) sucrose is moving out of Tube 3, and water

BIOLOGY E/M – Continued

98. Why does the mass of Tube 1 remain relatively unchanged throughout the experiment?

    (A) the dialysis tubing in Tube 1 is defective and does not allow water to cross
    (B) there is no concentration gradient to drive the movement of sucrose
    (C) the dialysis tubing broke, allowing the tube contents to mix with the beaker contents
    (D) there is no concentration gradient to drive the movement of water
    (E) the experimenter failed to record the data properly

99. Which of the following graphs best illustrates the relationship between Tube 2 and Tube 3?

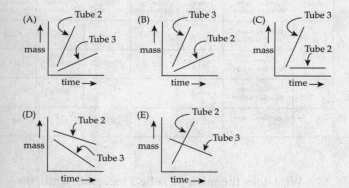

100. Cell membranes are also semipermeable, allowing water, but not other substances, to cross easily. A red blood cell placed in a 0.9% NaCl solution will neither swell nor shrivel. Based on this knowledge, and the information presented in Table 1, what would happen to a red blood cell placed in a 20% NaCl solution?

    (A) water would be drawn out of the cell and the cell would swell
    (B) water would be drawn into the cell and the cell would swell
    (C) water would be drawn out of the cell and the cell would shrivel
    (D) water would be drawn into the cell and the cell would shrivel
    (E) no change would occur to the cell

9

Explanations for the SAT II: Biology E/M Subject Test

Core Test

Questions 1-3

(A) mitochondria
(B) cytoplasm
(C) pyruvate
(D) lactic acid
(E) glucose

1. Location of cellular respiration in prokaryotes.

B is correct. Prokaryotes (bacteria, monerans) have no membrane bound organelles, so all reactions and processes must occur in the cytoplasm.

2. End product of anaerobic metabolism in muscle cells.

D is correct. When muscle cells run out of oxygen and switch to anaerobic metabolism (glycolysis only) to make ATP, the end product is lactic acid. Yeast can also switch to anaerobic metabolism; their end product is ethanol.

3. Location of glycolysis in eukaryotes.

B is correct. Eukaryotes possess organelles, and divide the location of their cellular processes among them. Glycolysis occurs in the cytoplasm, while the Krebs cycle and electron transport occur in the mitochondria.

Questions 4-6

(A) anaphase II
(B) metaphase I
(C) prophase II
(D) metaphase II
(E) prophase I

4. Stage of meiosis during which recombination of genetic material occurs.

E is correct. Recombination occurs when the homologous chromosomes are paired and crossing over can take place. This occurs during prophase I of meiosis.

5. Stage of meiosis during which pairs of homologous chromosomes align at the center of the cell.

B is correct. After prophase I, the homologous chromosomes remain paired and align at the center of the cell, on the metaphase plate. During metaphase II, the individual, *unpaired* chromosomes align at the cell center.

6. Stage of meiosis during which sister chromatids are separated.

A is correct. During meiosis, the chromosomes remain replicated (i.e. remain as two joined sister chromatids) for the entire first set of divisions. The whole point to the second set of meiotic divisions is to separate the sister chromatids. This takes place during anaphase II.

Questions 7-9
(A) reasoning/insight
(B) imprinting
(C) classical conditioning
(D) habituation
(E) instinct

7. A simple kind of learning involving loss of sensitivity to unimportant stimuli.

D is correct. Habituation involves becoming accustomed to certain stimuli that are not harmful or important. For example, if you walk down the hallway and a friend jumps out at you and you get scared, that is a normal reaction to a startling stimulus. However, if this happens every time you walk down the hallway, you get accustomed to it and no longer get startled. You have become habituated to the stimulus.

8. Geese recognize a ticking clock as "mother" if exposed to it during a critical period shortly after hatching.

B is correct. Some animals do not have an instinctive sense for who their mother is, and will bond with any object they are exposed to during a certain time period after their birth. The object imprints on their minds and thereafter, even if exposed to their real mother, they will still treat the object as mom.

9. Fish are given food at the same time as a tap on their glass bowl, and soon learn to approach when a tap sounds even in the absence of food.

C is correct. Conditioning involves the association of and response to one stimulus with a second, different stimulus. The best example is Ivan Pavlov's dogs. He rang a bell when he fed them, and the dogs salivated in response to the food. Soon, all he had to do was ring the bell, and the dogs would salivate, even in the absence of food.

Questions 10-12
(A) small intestine
(B) large intestine
(C) stomach
(D) esophagus
(E) mouth

10. Structure where most digestion and absorption of nutrients occurs.

A is correct. Most digestion and absorption occurs in the small intestine. A very small amount of digestion (starch only) takes place in the mouth, and a very small amount of digestion takes place in the stomach (acid hydrolysis of food and some protein digestion).

11. Structure where starch digestion first takes place.

E is correct. Saliva contains the enzyme amylase, which breaks down starch.

12. Structure with the lowest pH.

C is correct. Cells in the stomach secrete hydrochloric acid, which keeps the pH of the stomach around 1–2. The other regions of the digestive tract maintain a fairly neutral pH.

13. Homologous structures, which have similar underlying structures but may have different functions, are formed by

    (A) divergent evolution
    (B) speciation
    (C) segregation
    (D) convergent evolution
    (E) stabilizing selection

A is correct. Divergent evolution occurs when the same ancestral organism is placed into different environments, and must then adapt to function in these different environments. Thus the same original structures evolve separately, and diverge from one another. Examples of homologous structures are the arm of a man, the wing of a bat, and the flipper of a whale. All have the same basic bone structure, but vastly different functions. The opposite of divergent evolution is convergent evolution, where vastly different organisms are placed into the same environment, and must adapt to perform similar functions with different structures. Convergent evolution produces analogous structures, examples of which are the wings of bats, the wings of birds, and the wings of butterflies. Speciation is often the result of divergent evolution, not the cause of it.

14. Hemoglobin is a protein in red blood cells that binds and carries oxygen and some carbon dioxide. Its affinity for oxygen changes as blood travels from the lungs to the body tissues and back to the lungs again. One could expect hemoglobin to have

    (A) a high carbon dioxide affinity in the lungs and a low oxygen affinity in the tissues
    (B) a low carbon dioxide affinity in the lungs and a high oxygen affinity in the tissues
    (C) a high oxygen affinity in the lungs and a low oxygen affinity in the tissues
    (D) a low oxygen affinity in the lungs and a high oxygen affinity in the tissues
    (E) a high oxygen affinity in the lungs and a high carbon dioxide affinity in the lungs

C is correct. The job of the blood is to carry oxygen from the lungs, where it is plentiful, to the tissues, where it is not. Thus hemoglobin should have a high affinity for oxygen in the lungs (so it can bind oxygen) and a low oxygen affinity in the tissues (so it can release the oxygen where it is needed). The reverse is true for carbon dioxide. Hemoglobin has a high carbon dioxide affinity in the tissues, and a low carbon dioxide affinity in the lungs.

15. Which of the following RNA sequences would be transcribed from the DNA sequence ATGCCTAGGAC?

    (A) TACGGATCCTG
    (B) UAGCGAUCCUG
    (C) AUGCCUAGGAC
    (D) UACGGAUCCUG
    (E) GCAUUCGAAGU

D is correct. In RNA, the base thymine (T) is replaced with uracil (U), so choice A can be immediately eliminated. Further, A will always pair with U, and G will always pair with C. The only choice which has the bases paired correctly is choice D.

16. Arthropods can be characterized by all of the following EXCEPT

    (A) a hard exoskeleton
    (B) a water vascular system
    (C) jointed appendages
    (D) molting
    (E) segmented body

B is correct. Choices A, C, D, and E all describe characteristics of the phylum arthropoda. A water vascular system is a characteristic of the phylum echinodermata, the spiny skinned animals such as starfish and sea urchins. Their water vascular system ends in tube feet that play a role in locomotion and feeding.

17. Which of the following are functions of the kidney?

    I. filtration of blood to remove wastes
    II. blood pressure regulation
    III. pH regulation

    (A) I only
    (B) I and II only
    (C) I and III only
    (D) II and III only
    (E) I, II, and III

E is correct. The kidney's primary role is to filter blood to remove wastes (statement I is true), but it is also involved to a fair extent in blood pressure regulation (through renin and aldosterone, II is true) and in pH regulation (through excretion of hydrogen ion, III is true).

18. In chickens, the allele for long tail feathers (T) is dominant over the allele for short tail feathers (t). If a pure-breeding long-tailed chicken (TT) mates with a pure-breeding short-tailed chicken (tt), what percentage of their offspring (if mated with the correct genotype) could give rise to chickens with short tails?

    (A) 25%
    (B) 50%
    (C) 75%
    (D) 100%
    (E) unable to determine from the information given

D is correct. If a pure-breeding long-tailed chicken (TT) mates with a pure-breeding short-tailed chicken (tt), all of their offspring (the F1 generation) will have the genotype Tt (and have long tails). So all of them, if mated with the correct genotype (Tt or tt), could produce offspring with short tails.

19. All of the following could be considered density-dependent factors affecting population growth EXCEPT

    (A) limited nutrients
    (B) climate temperature
    (C) build-up of toxins
    (D) predation
    (E) limited water

B is correct. Density-dependent factors are those that get more significant as the size of the population increases. Limited nutrients and water, toxic waste build-up, and predation are all issues that are of greater concern to a large population than to a small one. Only choice B, climate temperature, is not more worrisome to a large group than to a small one. It will affect all populations equally, regardless of their size.

20. The best definition of a species is
    (A) a group of organisms that occupy the same niche
    (B) a population that works together to defend itself from predators
    (C) a group of organisms that can mate with each other
    (D) a population that preys on other populations
    (E) a population where all members benefits from the association in some way

C is correct. Two populations are considered separate species when they are so different from one another that they can no longer mate and produce viable offspring. Thus, organisms that can mate with each other must be of the same species.

21. Which of the following contains blood poor in oxygen?

    I. right ventricle
    II. pulmonary vein
    III. pulmonary artery

    (A) I only
    (B) II only
    (C) III only
    (D) I and II only
    (E) I and III only

E is correct. Blood that is poor in oxygen returns from the body to the right side of the heart (I is true), then travels through the pulmonary artery (III is true) to get to the lungs, where it picks up oxygen again. This oxygen-rich blood returns to the left side of the heart through the pulmonary vein (II is false), and is pumped back out to the body through the aorta.

22. An organism appears to be a segmented worm. Upon observation it is determined that the organism has a closed circulation, a mouth and an anus, and does NOT have an exoskeleton. The organism most likely belongs to the phylum
    (A) mollusca
    (B) annelida
    (C) echinodermata
    (D) arthropoda
    (E) chordata

B is correct. The characteristics described are those of the phylum annelida, the most common example of which is the earthworm. Mollusks have external shells (snails), echinoderms and arthropods have exoskeletons (starfish, crustaceans, insects), and chordates have endoskeletons, and in any case, are not worms.

23. Which of the following substances are produced by the light reactions of photosynthesis?
    (A) ATP and NADPH
    (B) ATP and glucose
    (C) NADH and glucose
    (D) ATP and NADH
    (E) NADPH and glucose

A is correct. The light reactions of photosynthesis use solar energy to power the production of ATP and NADPH (a reduced electron carrier). The ATP and NADPH (i.e. energy) produced during these reactions is used later during the Calvin cycle to fix carbon dioxide into carbohydrates.

24. Consider the following graph (see page 232) of substrate concentration vs. product formation. Assume enzyme concentration to be constant. Why does the graph level off at high substrate concentrations?

(A) all the enzyme is used up and product formation cannot occur without enzyme
(B) there is no more substrate to be converted into product
(C) substrate concentration exceeds enzyme concentration and all active sites are saturated
(D) the reaction has run to completion
(E) an inhibitor has been added and it has slowed down the rate of product formation

C is correct. When the concentration of substrate far exceeds the concentration of enzyme (remember, enzyme concentration is assumed to be constant), all the enzyme active sites are saturated with substrate, and the product is being formed at its maximum rate. The only way to increase product formation at this point is to increase the concentration of the enzyme. Note that enzymes should not be used up in the course of the reaction (A is wrong). Furthermore, product formation is still occurring, just at a stable rate (B and D are wrong). There is no reason to assume an inhibitor has been added.

25. A bird which feeds on both insects and berries would be classified as a

　I. primary consumer
　II. secondary consumer
　III. tertiary consumer

(A) I only
(B) II only
(C) III only
(D) I and II only
(E) II and III only

D is correct. Berries are plant products (i.e. primary producers), so any organism that eats berries is a primary consumer, or an herbivore (I is true). Secondary consumers, carnivores, and omnivores (e.g. birds) eat primary consumers (e.g. bugs), so II is also true. Tertiary consumers are carnivores (e.g. cats) that eat other carnivores (e.g. birds, secondary consumers). III is false.

26. Which of the following chemical formulas could represent a carbohydrate?

(A) $C_6H_6O_6$
(B) $C_3H_6O_3$
(C) $C_6H_{12}O_3$
(D) $C_5H_{10}O_{10}$
(E) CH_2O_4

B is correct. Carbohydrates have the general molecular formula $C_nH_{2n}O_n$, for example, glucose, $C_6H_{12}O_6$. The only formula that fits this rule is choice B.

27. A population of birds lives in an area with many insects upon which they feed. The insects live inside trees, burrowing into the bark. Over many hundreds of years, the average beak size in the bird population has increased. This is due to

   (A) increased fitness of the birds, leading to speciation
   (B) decreased fitness of the insects, allowing the birds to catch them more easily
   (C) increased fitness of large-beaked birds, leading to natural selection
   (D) decreased fitness of small-beaked birds, leading to speciation
   (E) random mutation and genetic recombination

C is correct. Speciation has not occurred, only a change in the characteristic of the birds, so choices A and D can be eliminated. Any change in the fitness of the insects would change the characteristics of the insect population, not the bird population (B is wrong), and random mutation would not produce a specific, directed effect; (E is wrong). Birds with large beaks had greater fitness because they could more easily obtain food, they had an advantage over birds with smaller beaks, which died out as time passed.

28. The location on an enzyme where substrate binds is called the

   (A) binding site
   (B) reaction center
   (C) allosteric site
   (D) lock-and-key model
   (E) active site

E is correct. The other choices do not describe substrate binding sites on an enzyme.

29. Human cells maintain concentration gradients across their plasma membranes, such that there is a high sodium concentration outside the cell and a high potassium concentration inside the cell. Within the cell membrane are potassium "leak" channels; channels which are open all the time and through which potassium constantly exits the cell. This manner of transport across the plasma membrane is known as

   (A) simple diffusion
   (B) exocytosis
   (C) active transport
   (D) facilitated diffusion
   (E) secretion

D is correct. Any time a concentration gradient is being used to drive movement of something across the membrane, the process must be a form of diffusion (B, C, and E can be eliminated). Simple diffusion is reserved for those molecules that are lipid soluble and can cross the membrane without any help, such as oxygen and carbon dioxide (A is wrong). Facilitated diffusion is for those molecules that need help. In other words, a strong gradient exists to drive movement, but the molecules are not lipid soluble and cannot cross the membrane without a pathway, like a channel or carrier protein.

30. The role of decomposers in the nitrogen cycle is to
    (A) fix atmospheric nitrogen into ammonia
    (B) incorporate nitrogen into amino acids and organic compounds
    (C) convert ammonia to nitrate, which can then be absorbed by plants
    (D) denitrify ammonia, thus returning nitrogen to the atmosphere
    (E) release ammonia from organic compounds, thus returning it to the soil

E is correct. Decomposers take organic material and break it down into its individual compounds, thus returning these compounds to the earth. The other processes listed are carried out by other organisms: nitrogen fixing bacteria (choice A), heterotrophs and autotrophs (choice B), other soil bacteria (choices C and D).

31. All of the following are true about the endocrine system EXCEPT
    (A) it relies on chemical messengers that travel through the bloodstream
    (B) it is a control system that has extremely rapid effects on the body
    (C) the hormones affect only certain "target" organs
    (D) it is involved in maintaining body homeostasis
    (E) its organs secrete hormones directly into the bloodstream, rather than through ducts

B is correct. The endocrine system is a body control system, but it is NOT rapid. The fastest hormone in the body is adrenaline, and even that takes a few seconds, compared to the nervous system's milliseconds. Most hormones operate in the minutes to hours range. The other choices regarding the endocrine system are all true.

32. Two organisms live in close association with one another. One organism is helped by the association, while the other is neither helped nor harmed. Which of the following terms best describes this relationship?
    (A) mutualism
    (B) commensalism
    (C) symbiosis
    (D) parasitism
    (E) predator-prey relationship

B is correct. This symbiotic relationship describes commensalism. In mutualism, both partners benefit, in parasitism and predator-prey relationships one partner benefits while the other is harmed, and symbiosis is a general term used to describe close living arrangements.

33. Cardiac output (the amount of blood pumped out of the heart in one minute) and blood pressure are directly proportional. Which of the following graphs (see page 234) best depicts the relationship between cardiac output and blood pressure?

D is correct. Relationships that are directly proportional have linear graphs with a positive slope.

Questions 34-36 refer to the diagram on page 234.

34. Location where male haploid cells are produced.
    (A) 1
    (B) 2
    (C) 3
    (D) 6
    (E) 8

B is correct. Male haploid cells (pollen grains, microspores) are produced on the anther (#2 in the diagram), which is at the tip of the filament (#3 in the diagram).

35. Sticky structure where pollen grains can attach and germinate.
    (A) 1
    (B) 2
    (C) 4
    (D) 6
    (E) 8

A is correct. Pollen grains stick to the stigma (#1 in the diagram), which is supported by the style (#4 in the diagram).

36. Structure which, when fertilized, develops into fruit.
    (A) 1
    (B) 2
    (C) 5
    (D) 6
    (E) 8

E is correct. The pollen fertilizes the ovule (female haploid cells, megaspores, #6 in the diagram), once fertilized, the ovary (#8 in the diagram) develops into a fruit.

37. A plant growing on the shady side of a building bends around the corner of the building toward the sunlight. This is an example of

    (A) negative geotropism
    (B) negative phototropism
    (C) positive phototropism
    (D) positive hydrotropism
    (E) negative hydrotropism

C is correct. The prefix photo refers to light, since the plant is growing toward light we can eliminate choices A, D, and E. *Positive* means growing toward, so choice B can be eliminated.

38. The stem and leaves of the plant grow up, away from the soil. This is an example of

    (A) negative geotropism
    (B) positive geotropism
    (C) negative phototropism
    (D) positive hydrotropism
    (E) negative hydrotropism

A is correct. The prefix geo- refers to the earth, or soil, so we can eliminate choices C, D, and E. Since the plant is growing away from the earth, this is a negative tropism and we can eliminate choice B.

Questions 39-43 refer to the diagram on page 235.

39. The hormone labeled as "X" in the diagram is often used in over-the-counter diagnostic tests to determine when ovulation has occurred. This hormone is

   (A) estrogen
   (B) progesterone
   (C) FSH
   (D) LH
   (E) testosterone

D is correct. A surge in LH is what causes ovulation and is measured in the ovulation prediction kits. Estrogen and progesterone affect the uterus, not the ovary, and FSH causes development of a follicle (A, B and C are wrong). Testosterone is a male hormone (E is wrong).

40. Based on the peak levels of hormone "X", on what day of the cycle is ovulation most likely to occur?

   (A) day 21
   (B) day 14
   (C) day 12
   (D) day 25
   (E) day 28

B is correct. Since we know hormone X is peaking at ovulation time, a quick look at the graph shows hormone X peaking at about day 14 of the cycle.

41. The hormone labeled as "Y" in the diagram is

   (A) progesterone, secreted by the corpus luteum after ovulation has occurred
   (B) progesterone, secreted by the ovary after ovulation has occurred
   (C) estrogen, secreted by the corpus luteum after ovulation has occurred
   (D) estrogen, secreted by the ovary after ovulation has occurred
   (E) estrogen, secreted by the follicle before ovulation occurs

A is correct. The rise in hormone Y occurs after ovulation (choice E is wrong) and coincides with formation of the corpus luteum (choices B and D are wrong). The primary hormone secreted by the corpus luteum is progesterone.

42. Immediately after fertilization, the zygote begins to undergo rapid cell division. This process is known as

   (A) blastulation
   (B) gastrulation
   (C) neurulation
   (D) implantation
   (E) cleavage

E is correct. Rapid cell division after fertilization is known as cleavage. Blastulation is the formation of a hollow ball of cells (A is wrong), gastrulation is formation of the three primary germ layers (B is wrong), neurulation is development of the nervous system (C is wrong), and implantation is when the morula (solid ball of cells) burrows into the uterine lining (D is wrong).

43. From which of the primary germ layers does the nervous system develop?

    (A) endoderm
    (B) mesoderm
    (C) ectoderm
    (D) enteroderm
    (E) epidermis

C is correct. The ectoderm forms the skin, hair, nails, mouth lining, and nervous system. The mesoderm forms muscle, bone, blood vessels, and organs (B is wrong). The endoderm forms inner linings and glands (A is wrong). Enteroderm and epidermis are not primary germ layers.

44. The process which has taken place can best be described as

    (A) progression
    (B) succession
    (C) evolution
    (D) habitation
    (E) colonization

B is correct. The development of a thriving ecosystem from a barren area is known as succession. Note that evolution usually has a much longer time frame than succession.

45. The stable community of deciduous trees and animals is known as the

    (A) final community
    (B) climax community
    (C) apex community
    (D) summit community
    (E) composite community

B is correct. The climax community is the final, stable community in succession.

46. Usually the first organisms to colonize rocky areas are lichen. These are known as the

    (A) primary community
    (B) starter community
    (C) colony organisms
    (D) pioneer organisms
    (E) settler organisms

D is correct. The first organisms to colonize a barren area are known as the pioneer organisms.

Questions 47-50 refer to the experiment on page 236.

47. Which of the following substances could be classified as a diuretic?

    I. caffeine
    II. sodium
    III. water

    (A) I only
    (B) II only
    (C) I and II only
    (D) II and III only
    (E) I, II, and III

A is correct. Diuretics help eliminate water (i.e. increases urine production) from the body. From the data tables, the only substance that increases urine production significantly is caffeine.

48. Which graph (see page 236) best represents the change in urine volume when ingesting caffeine?

A is correct. Again, from the data tables, there is a directly proportional (i.e. linear) relationship between the amount of caffeine ingested and the volume of urine produced. As caffeine consumption increases, so does urine volume. The only graph that shows this relationship is choice A.

49. The purpose of ingesting the plain water was to

(A) rehydrate the volunteers
(B) dissolve the substances
(C) act as a control
(D) flush out the kidneys
(E) act as a positive test substance

C is correct. Since the caffeine and the sodium chloride were dissolved in water, plain water was consumed as a control, to make sure the effects seen were due to the added substances, and not the water.

50. Based on the results in Table II, if a volunteer were to ingest 4.5 g sodium chloride dissolved in 100 ml water, what would be the approximate predicted urine volume collected after one hour?

(A) 20 ml
(B) 30 ml
(C) 40 ml
(D) 50 ml
(E) 60 ml

C is correct. From Table 2, an increase in sodium chloride of 0.9 g results in a decrease in urine volume of approximately 40 ml. When 3.6 g sodium chloride are ingested, 82 ml urine is produced, thus if 4.5 g sodium chloride were to be ingested, the expected urine volume would be 40 ml less, approximately 40 ml.

51. Which of the following best explains the reason that Mark, Mike, Molly and Mary do not display symptoms of hemophilia, even though their father, John, is a hemophiliac?

(A) hemophilia is an X-linked disorder, and John can only pass on his Y chromosome
(B) hemophilia is an X-linked disorder, and even though Molly and Mary received a hemophiliac X chromosome from John, Jane gave them a normal X chromosome
(C) hemophilia is a Y-linked disorder, and therefore cannot be displayed in females
(D) hemophilia is a Y-linked disorder, and Mark and Mike must have received an X chromosome from John
(E) hemophilia is an X-linked disorder, and even though Mark and Mike received a hemophiliac X chromosome from John, Jane gave them a normal X chromosome

B is correct. Out of this family, the only members that express this condition are males. This is a tip off for X-linked disorders, which are more common in males because they have only a single X chromosome. John's genotype is YXh. He passed his Y chromosome to Mark and Mike; they also received a normal X from Jane, thus they do not have hemophilia nor can they pass it on to their kids. Molly and Mary received Xh from John, but also received a normal X from Jane, thus they are carriers of hemophilia, but do not display its symptoms.

52. If one of Mike's daughters marries a normal man, what is the probability that one of their children will display symptoms of hemophilia?

    (A) 0%
    (B) 25%
    (C) 50%
    (D) 75%
    (E) 100%

A is correct. Since Mike does not carry the gene for hemophilia (see above), he cannot pass it on to his children, and they in turn cannot pass it on to their children.

53. Which of the following individuals are heterozygous for hemophilia?

    (A) John, Mark, and Mike
    (B) Mark, Mike, Molly, and Mary
    (C) John and Jane
    (D) Molly and Mary
    (E) Mark and Mike

D is correct. Mark and Mike do not carry the gene for hemophilia (see solution to 51 above), thus we can eliminate choices A, B, and E. Jane is normal, thus choice C is eliminated as well.

Questions 54-57 refer to the table on page 237.

54. Based on the above information, which of the following is most likely substance B?

    (A) calcitonin
    (B) insulin
    (C) parathyroid hormone
    (D) glucagon
    (E) aldosterone

C is correct. From Table 1, substance B caused an increase in blood calcium levels, which is the effect that parathyroid hormone has on the body.

55. Based on the above information, which of the following is most likely substance A?

    (A) glucagon
    (B) aldosterone
    (C) calcitonin
    (D) parathyroid hormone
    (E) insulin

E is correct. Again, from Table 1, substance A caused a decrease in blood glucose levels, which is the effect that insulin has on the body.

56. Under what conditions might substance D be released normally?

(A) soon after a meal
(B) when blood pressure is low
(C) in between meals
(D) when there has been limited intake of dietary calcium
(E) when dietary calcium is in excess

B is correct. Substance D causes an increase in blood sodium, which is the effect aldosterone has on the body. Aldosterone is released when blood pressure is low, since excess sodium will have the effect of causing water retention, which will increase blood volume, which will increase blood pressure. (Note: even if you did not know this, you should have been able to eliminate the other choices.)

57. All of the following changes in variable values are significant EXCEPT

(A) the change in serum glucose when substance A is injected
(B) the change in serum Na⁺ when substance D is injected
(C) the change in serum Ca⁺⁺ when substance B is injected
(D) the change in serum glucose when substance C is injected
(E) the change in serum Na⁺ when substance B is injected

E is correct. The change in serum sodium after injection of Substance B is insignificant. All other choices cause significant change from the baseline values of the variables being measured.

58. Based on the information in Table 1 (see page 238), which of the following is the most likely classification of cell Type A?

(A) fungi
(B) plant
(C) bacteria
(D) animal
(E) protist

C is correct. Cell Type A has no nucleus. The only organisms that do not have nuclei are bacteria (kingdom Monera).

59. Which of the following equations is cell Type C able to run?

I. $C_6H_{12}O_6 + 6\,O_2 \rightarrow 6\,CO_2 + 6\,H_2O + ATP$
II. $H_2O + light \rightarrow O_2 + ATP + NADPH$
III. $6\,CO_2 + 6\,H_2O + ATP + NADPH \rightarrow C_6H_{12}O_6$

(A) I only
(B) II only
(C) I and III only
(D) II and III only
(E) I, II, and III

E is correct. Cell Type C has a nucleus, a cell wall, and chloroplasts, and is most likely a plant. Equations II and III are the equations for photosynthesis, and would occur in plants, and equation I is the equation for cellular respiration, which also occurs in plants.

60. Consider Figure 1 (see page 238). Which of the following statements best describes the reason for the difference between the curves for cell Type B and cell Type C?
    (A) cell Type B is unable to survive in the presence of oxygen, while cell Type C can ferment
    (B) the products of fermentation in cell Type C are toxic to the cells and they are dying
    (C) cell Type B is an obligate aerobe while cell Type C is able to ferment
    (D) cell Type B is a facultative anaerobe, while cell Type C is an obligate aerobe
    (E) cell Type C is an obligate aerobe, while cell Type B is an obligate anaerobe

D is correct. At Time 1 the oxygen was removed from the cultures and cell Type C died. Clearly it is an obligate aerobe. So we can eliminate choices A, B, and C. Since cell Type B was growing well in the presence of oxygen, it cannot be an obligate anaerobe, thus choice E is eliminated. Cell Type B must be a facultative anaerobe, using oxygen when it is available, and fermenting when oxygen is not available. The decrease in growth of cell Type B after Time 1 is most likely due to the fact that less energy is produced during fermentation than during aerobic metabolism.

Questions 61-64
(A) tundra
(B) taiga
(C) chaparral
(D) deciduous forest
(E) desert

61. The driest of all terrestrial biomes, characterized by low and unpredictable precipitation.

E is correct. These are the characteristics of desert.

62. Coniferous forests, characterized by long, cold winters and short, wet summers.

B is correct. These are the characteristics of taiga.

63. Coastal areas characterized by mild, rainy winters and long, hot, dry summers.

C is correct. These are the characteristics of chaparral.

64. Northern areas, characterized by permafrost, extremely cold temperatures, and few trees.

A is correct. These are the characteristics of tundra.

65. Plants that have true roots, stems, and leaves, as well as flowers and seeds enclosed in fruit are classified as

    (A) bryophytes
    (B) tracheophytes
    (C) gymnosperms
    (D) angiosperms
    (E) endosperms

D is correct. Flowering plants are angiosperms. Gymnosperms are conifers (naked seed plants, C is wrong), bryophytes are mosses (A is wrong), and tracheophytes are non-seed producing plants (ferns, B is wrong). Endosperm is not a classification for plants.

66. Which of the following indicates that animals have internal biological clocks?

    (A) a mouse kept in constant darkness shows a daily rhythm of activity
    (B) a rooster crows whenever the sun rises in both winter and summer
    (C) an owl kept in constant light drifts away from a 24-hour cycle
    (D) some species of birds can sense fluctuations in the Earth's magnetic field
    (E) a squirrel whose night and day are reversed artificially soon adapts to its new schedule

A is correct. If an organism's environment remains absolutely constant, and that organism still exhibits regular rhythms of activity, there must be some internal clock that keeps it on schedule (C and E are wrong). Roosters vary the time of their crow as the sun varies the time it rises (B is wrong). The magnetic field has nothing to do with internal clocks.

67. Which of the following correctly lists the phylogenic hierarchy?

    (A) kingdom, phylum, family, class, order, genus, species
    (B) phylum, family, order, class, kingdom, species, genus
    (C) kingdom, family, order, class, phylum, genus, species
    (D) kingdom, phylum, class, order, family, genus, species
    (E) family, kingdom, order, phylum, genus, class, species

D is correct. Remember: King Philip Came Over From Germany—so?, or make up your own!

68. A rattlesnake would be classified as a

    (A) tertiary consumer and a heterotroph
    (B) secondary consumer and an autotroph
    (C) producer and an autotroph
    (D) producer and a heterotroph
    (E) primary consumer and a heterotroph

A is correct. Rattlesnakes are clearly heterotrophs (only photosynthetic organisms are autotrophs) so we can eliminate choices B and C. Rattlesnakes are carnivores, not producers (D is wrong) or primary consumers (herbivores, E is wrong).

69. At some point in their development, chordates possess all of the following EXCEPT

   (A) a dorsal hollow nerve cord
   (B) a notochord
   (C) gill slits
   (D) postnatal tail
   (E) an exoskeleton

E is correct. The only characteristic not possessed by chordates is an exoskeleton.

<u>Questions 70-73</u> refer to the diagrams on page 240.

70. What is the average beak length (in cm) of the birds in Figure 1?

   (A) 30 cm
   (B) 15 cm
   (C) 5 cm
   (D) 3 cm
   (E) 1 cm

D is correct. The question asks for the average beak length in cm, but the graph gives it in mm. Average beak length is 30 mm. 10 mm = 1 cm, therefore, 30 mm = 3 cm.

71. What is the most likely reason for the difference in distribution of beak lengths between the data plotted in Figure 1 and the data plotted in Figure 2?

   (A) all birds with beaks of 30 mm flew to a new island over the 200-year time span
   (B) birds with beaks of 30 mm were selected against
   (C) predators consumed birds with beaks of 40 mm
   (D) predators consumed birds with beaks of 20 mm
   (E) birds with beaks of 30 mm were selected for

B is correct. Clearly the birds with 30 mm beaks were not surviving too well. There is no reason to assume they flew to another island, remember, they are on a remote, isolated island. There may not be another island near enough to fly to (A is wrong). If predators consumed birds with 20 mm or 40 mm beaks, they would not be the prevalent populations in Figure 2 (C and D are wrong), and if birds with 30 mm beaks were selected for, the population would not have been divided (E is wrong).

72. Suppose that a researcher studying Population B found that birds with beaks of 20 mm were unable to mate with birds that had 40 mm beaks. These two groups of birds would now be classified as

   (A) occupying different niches
   (B) separate species
   (C) competitors
   (D) predators
   (E) separate populations

B is correct. The defining characteristic for speciation is an inability to interbreed.

73. How might beak length in the bird population change after another 200-year time span?

   (A) the average beak length would return to 30 mm
   (B) the average beak length would shift to 40 mm
   (C) the average beak length would shift to 20 mm
   (D) the differences in beak length would be more pronounced
   (E) it is not possible to determine how beak length might change

E is correct. Just because we have some information about how the population changed in the last 200 years, this doesn't tell us how it may change in the next 200 years. It would depend on how the environment changed during that time period.

74. The effect of acid rain on fish size is best represented by which of the following graphs (see page 241)?

D is correct. As the acidity increases (pH goes down), the average mass of the fish decreases. However, it does not decrease linearly, rather it stays constant for a while, then gradually drops off, then rapidly drops off as acidity becomes severe. The best representation of this is choice D.

75. The relationship between mycorrhizal fungi and plants can best be described as one where

   (A) one partner benefits from the association and the other partner is harmed
   (B) one partner benefits from the association and the other partner is neither harmed nor helped
   (C) one partner preys upon the other partner
   (D) both partners benefit from the association
   (E) neither partner benefits from the association

D is correct. The plant benefits by easier availability of water and phosphorus, the fungi benefit by receiving amino acids and sugars. Another term for this type of relationship is mutualism.

76. If the pH of the soil were 7.0, what would be the effect on the mycorrhizal fungi and plants?

   (A) the fungi would survive but the plant would be harmed
   (B) the fungi would be harmed but the plant would survive
   (C) the fungi would be slightly harmed and the plant would be slightly harmed
   (D) neither the fungi nor the plant would survive
   (E) neither the fungi nor the plant would be harmed

E is correct. pH 7.0 is neutral, so neither the plant nor the fungi would be harmed.

77. What might be the best strategy to prevent ecological damage due to acid rain?

   (A) stock the lakes with bigger fish so that they can resist the effects of the acid better
   (B) reduce the amount of fossil fuels that are burned
   (C) supply plants with excess phosphorus and water
   (D) supply fungi with excess sugars and amino acids
   (E) only fish when it is sunny

B is correct. The best way to prevent damage from acid rain would be to prevent its formation by reducing the burning of the fossil fuels that cause it. There is no guarantee that bigger fish will resist the acidity any better than smaller fish (A is wrong), supplying plants with excess phosphorus will not help them take it up any easier (C is wrong), supplying fungi with sugars and amino acids will not help them overcome the effects of acid soil (D is wrong), and only fishing when it's sunny will protect you from acid rain, but not the environment!

78. Fungi are classified as

   (A) prokaryotic decomposers
   (B) eukaryotic producers
   (C) eukaryotic decomposers
   (D) eukaryotic autotrophs
   (E) prokaryotic consumers

C is correct. Fungi are eukaryotes (A and E are eliminated), and they are not photosynthetic (B and D are eliminated).

79. The data in Figure 2 (see page 242) indicate that

   (A) *P. aurelia* is preying on *P. caudata*
   (B) *P. aurelia* is a better competitor than *P. caudata*
   (C) *P. aurelia* and *P. caudata* are in a symbiotic relationship
   (D) *P. aurelia* is a parasite of *P. caudata*
   (E) *P. aurelia* grew better when combined with *P. caudata* that it did when grown alone

B is correct. Clearly *P. aurelia* can compete better and get more food that *P. caudata*, so it will grow while *P. caudata* is competed to extinction. A is highly unlikely, since the food source the paramecia prefer is bacteria, not each other. This is not a symbiotic relationship, but a competitive one (C and D are eliminated), and the data contradict E.

80. Paramecia are members of the kingdom

   (A) fungi
   (B) animalia
   (C) monera
   (D) protista
   (E) plantae

D is correct. Paramecia are single-celled eukaryotes, members of kingdom Protista.

81. All of the following is true about RNA EXCEPT

    (A) it is single stranded
    (B) its bases are adenine, thymine, guanine, and uracil
    (C) it has a sugar-phosphate backbone
    (D) its sugar is ribose
    (E) it is found in the both the nucleus and the cytoplasm of the cell

B is correct. RNA bases do not include thymine; they are adenine, guanine, cytosine, and uracil. All other statements about RNA are correct.

82. The function of the Golgi apparatus is to

    (A) package and store proteins for secretion
    (B) synthesize proteins
    (C) function in cellular respiration
    (D) help the cell expel waste
    (E) digest foreign substances

A is correct. Ribosomes synthesize protein (B is wrong), mitochondria function in respiration (C is wrong), vacuoles help expel waste (D is wrong), and lysosomes function in digestion (E is wrong).

83. A eukaryotic cell that has a cell wall but lacks chloroplasts would be classified as a

    (A) moneran
    (B) chordate
    (C) plant
    (D) fungus
    (E) bacteria

D is correct. Choices A and E are prokaryotic and can be eliminated. Chordates have no cell walls (B is wrong), and plants have chloroplasts (C is wrong).

84. All of the following could give rise to new species EXCEPT

    (A) variations in antler size between male and female reindeer
    (B) an earthquake that physically separates a population of lizards into two separate groups
    (C) divergent evolution
    (D) evolution of a population of cats such that they can no longer mate with their ancestors
    (E) a massive flood that separates a population of frogs onto opposite sides of a large lake

A is correct. The defining characteristic for speciation is an inability to interbreed. Choices B, C, D, and E could all ultimately produce two different populations that lack the ability to mate. Choice A would not lead to an inability to mate.

85. The nucleic acid that is translated to make a protein is

    (A) rRNA
    (B) DNA
    (C) hnRNA
    (D) tRNA
    (E) mRNA

E is correct. rRNA helps form the ribosome, DNA is the genetic material of the cell, hnRNA is unspliced mRNA, and tRNA transfers amino acids to ribosomes.

86. Which of the following groups have the most in common with one another?

    (A) members of the same kingdom
    (B) members of the same genus
    (C) members of the same phylum
    (D) members of the same class
    (E) members of the same family

B is correct. The members that have the most in common with one another are the members near the bottom of the hierarchy. Of the choices given, genus is the closest to the bottom of the hierarchy.

87. Which of the following individuals is the LEAST fit in evolutionary terms?

    (A) a 45-year old male with a terminal disease who has fathered three children
    (B) a 20-year old man who has fathered one child
    (C) a 35-year old woman with four children
    (D) a healthy 4-year old child
    (E) a 25-year old woman with one child, who has had a tubal ligation to prevent future pregnancies

D is correct. Anyone who has produced offspring has demonstrated their fitness. Regardless of how healthy a child is, he has not yet produced offspring to prove his fitness.

88. Referring to Figure 1 (see page 243), what is the genotype of Colony 3?

    (A) arg^+, leu^+, pro^+
    (B) arg^+, leu^-, pro^+
    (C) arg^+, leu^+, pro^-
    (D) arg^-, leu^-, pro^+
    (E) arg^-, leu^-, pro^-

C is correct. The only plate that Colony 3 cannot grow on is plate C, which lacks proline. So Colony 3 requires proline to grow and is a proline auxotroph (pro()). This eliminates choices A, B, and D. Choice E is incorrect because Colony 3 does not require arginine or leucine to grow, it can grow just fine in the absence of these amino acids, as is indicated on plates A and B.

89. Is Colony 1 (see page 243) an auxotroph?

    (A) Yes, it is able to grow in the presence of the three amino acids being tested.
    (B) Yes, it can only grow if glucose is present.
    (C) No, it is able to grow in the absence of glucose.
    (D) No, it is able to grow in the absence of any additional amino acids.
    (E) The data available are insufficient to determine the answer.

D is correct. Colony 1 can grow on any of the plates, thus it does not require any additional amino acids. Auxotrophs require additional supplements to their growth media.

90. Which structures could be observed in a sample of Colony 2 (see page 243)?

   I. nuclei
   II. ribosomes
   III. mitochondria

   (A) I only
   (B) I and III only
   (C) I, II, and III
   (D) II and III only
   (E) II only

E is correct. Since these are bacterial colonies, they would not have any membrane-bound organelles, so no nuclei or mitochondria. Bacteria do have ribosomes to synthesize proteins.

91. If a liquid culture medium containing glucose, leucine, and proline was inoculated with Colony 4, would bacterial growth be observed?

   (A) No, Colony 4 is an arginine auxotroph (arg⁻).
   (B) No, Colony 4 cannot grow in the presence of leucine.
   (C) Yes, Colony 4's genotype is leu⁻, pro⁻.
   (D) Yes, Colony 4 requires only glucose to grow.
   (E) The data available are insufficient to make a prediction.

A is correct. Colony 4 cannot grow in the absence of arginine as is evidenced by plate B. Thus, since the liquid media does not contain arginine, no bacterial growth would be observed.

92. What is the most likely reason for the clearing (the plaque) in the lawn of bacteria in the second experiment?

   (A) The unknown organism is bacterial Colony 2, and these bacteria are eating the bacteria from Colony 1 forming the lawn.
   (B) The unknown organism is a virus that is infecting the bacteria and causing them to lyse (killing them).
   (C) The drop placed in the center of the lawn contained a strong acid that destroyed the bacteria at that spot.
   (D) Bacteria are very delicate and the disturbance caused them to die.
   (E) The unknown organism began producing threonine, which is toxic to Colony 1.

B is correct. Clear spots on lawns of bacteria are due to infection by viruses that cause lysis of the bacteria. Even if this was not obvious to you, you should have been able to eliminate the other choices. There is no reason to assume Colony 2 would prey on Colony 1 (A is wrong), strong acid would lyse and destroy the bacteria immediately, not after 24 hours (C is wrong), bacteria are not delicate, they can grow just about anywhere, under any conditions (D is wrong), and there is no data to support the fact that threonine may be toxic to the bacteria, or that the unknown organism was producing it.

Questions 93-96 refer to the table on page 244.

93. The wings of the moths and the wings of the birds are both used for flight (similar functions), however their underlying structures are very different. Moth wings and bird wings are thus classified as

    (A) homologous structures
    (B) autologous structures
    (C) divergent structures
    (D) analogous structures
    (E) emergent structures

D is correct. Structures with similar functions, but different underlying structures are the result of vastly different organisms being placed into similar environments and having to adapt to the same stresses with different starting materials. They are termed analogous structures and are the result of convergent evolution. (See also #13 explanation.)

94. What is the most likely explanation for the shift in the percentage of black moths in the population?

    (A) the white moths no longer blended with the color of the tree bark, and thus were selected for
    (B) the black moths blended better with the color of the tree bark, and thus were selected for
    (C) the black moths blended better with the color of the tree bark, and thus were selected against
    (D) the white moths blended better with the color of the tree bark, and thus were selected against
    (E) the black moths did not blend with the color of the tree bark, and thus were selected against

B is correct. Since the percentage of black moths is increasing, they must be selected for. This eliminates C and E. Since the percentage of white moths is decreasing, they must be selected against, eliminating choice A. Choice D is wrong because white moths would not blend better against dark tree bark.

95. If a seed from one of the trees was planted in an area far from the steel mill, what color would the bark of the tree be?

    (A) black, since the parent tree had black bark
    (B) white, since the gene causing black bark was mutated due to environmental pollution
    (C) black, since the gene causing white bark was mutated due to environmental pollution
    (D) white, since the black bark was an acquired characteristic and is therefore not passed on to progeny
    (E) the color of the bark is not able to be determined

D is correct. The original parent had white bark. The change in bark color is due to an accumulation of soot on the tree. This is an acquired characteristic and would not be passed on to offspring. Seedlings that grew far from the plant would not be exposed to soot in the air and would not experience discoloration of their bark.

96. Birds track their prey visually, while bats rely on sonar to locate their food. If the bird population was replaced with a bat population in 1940, what would be the ratio of white moths to black moths?

    (A) 95% white, 5% black
    (B) 80% white, 20% black
    (C) 50% white, 50% black
    (D) 20% white, 80% black
    (E) 5% white, 90% black

C is correct. The reason the percentage of black moths increased was because they were no longer visible against the now darkened bark of the trees. Since the white moths were more easily seen by the birds, their numbers declined. However, since bats rely on sonar to locate prey instead of vision, darker coloring would not give the moths any advantage, and the population percentages would stay at the point they were at when the birds were replaced by bats, a fifty-fifty split.

Questions 97-100 refer to the table on page 245.

97. Why does the mass of Tube 3 increase, while the mass of Tube 4 decreases?

    (A) water is moving into Tube 3, and sucrose is moving into Tube 4
    (B) water is moving into Tube 4, and sucrose is moving into Tube 3
    (C) water is moving into Tube 3, and water is moving out of Tube 4
    (D) sucrose is moving into Tube 3, and sucrose is moving out of Tube 4
    (E) sucrose is moving out of Tube 3, and water is moving out of Tube 4

C is correct. Sucrose cannot cross the dialysis membrane, so it cannot cause any effect on the mass of the tubes. This eliminates choices A, B, D, and E.

98. Why does the mass of Tube 1 remain relatively unchanged throughout the experiment?

    (A) the dialysis tubing in Tube 1 is defective and does not allow water to cross
    (B) there is no concentration gradient to drive the movement of sucrose
    (C) the dialysis tubing broke, allowing the tube contents to mix with the beaker contents
    (D) there is no concentration gradient to drive the movement of water
    (E) the experimenter failed to record the data properly

D is correct. Since movement across the membrane relies strictly on concentration gradients, the fact that there is no gradient in Tube 1 would prevent the movement of water into or out of the tube. Thus there would be no change in mass.

99. Which of the following graphs best illustrates the relationship between Tube 2 and Tube 3?

B is correct. The gradient in Tube 3 is much larger than the gradient in Tube 2, so water would be expected to enter more rapidly. This is confirmed by the data in Table 1. A linear graph should show two lines with positive slopes, and the slope of Tube 3's line should be greater than the slope of Tube 2's line.

100. Cell membranes are also semipermeable, allowing water, but not other substances, to cross easily. A red blood cell placed in a 0.9% NaCl solution will neither swell nor shrivel. Based on this knowledge, and the information presented in Table 1, what would happen to a red blood cell placed in a 20% NaCl solution?
    (A) water would be drawn out of the cell and the cell would swell
    (B) water would be drawn into the cell and the cell would swell
    (C) water would be drawn out of the cell and the cell would shrivel
    (D) water would be drawn into the cell and the cell would shrivel
    (E) no change would occur to the cell

C is correct. The fact that the cell does not swell or shrivel in 0.9 % NaCl implies that there is no concentration gradient. A cell in a 20% NaCl solution would experience similar stresses to a dialysis tube filled with water sitting in a beaker filled with a more concentrated solution, such as Tube 4. The data indicate that Tube 4 lost mass, so water exited the tube, and the same would happen to the red blood cell.

10
Answers to In-Chapter Questions

CHAPTER 3: ANSWERS

QUESTIONS FROM PAGE 24

- H_2O [■ is   ❏ is not] a *compound*.
- Cl_2 [❏ is   ■ is not] a *compound*.
- Cl_2 [■ is   ❏ is not] a *molecule*.

QUESTIONS FROM PAGE 24

- Water (H_2O) is an [❏ organic   ■ inorganic] compound.
- Methane (CH_3) is an [■ organic   ❏ inorganic] compound.

QUESTIONS FROM PAGE 27

- A protein represents a chain of __amino__ __acids__.
- The bond that links two amino acids together is called a __peptide__ bond.
- Because proteins are, essentially, chains of amino acids linked together by __peptide__ bonds, a protein might also be called a __polypeptide__.

QUESTIONS FROM PAGE 27

- The process in which two amino acids are separated involves the breakage of a __peptide__ bond.
- The breakage of a peptide bond requires the addition of __water__ and is given the name __hydrolysis__.
- The disassembly of a polypeptide into its component amino acids involves the [■ addition   ❏ removal] of water, and is called __hydrolysis__.
- The assembly of a polypeptide from its component amino acids involves the [❏ addition   ■ removal] of water, and is called __condensation__.

QUESTIONS FROM PAGE 29

- Molecule A is called __glucose__.
- Molecule B is called __fructose__.
- The chemical formula for glucose is $C_6H_{12}O_6$.
- The chemical formula for fructose is $C_6H_{12}O_6$.
- Glucose and fructose [❏ are   ■ are not] identical molecules.
- Glucose and fructose [■ do   ❏ do not] have identical chemical formulas.

QUESTIONS FROM PAGE 30

- A molecule of maltose is formed from two molecules of _glucose_.
- A molecule of glucose and a molecule of fructose, both of which are _mono_saccharides, combine to form a molecule of _sucrose_, which is a _di_saccharide.
- The chemical formula for both glucose and fructose is $C_6H_{12}O_6$.
- The chemical formula for sucrose is $C_{12}H_{22}O_{11}$.

QUESTIONS FROM PAGE 31

- Large-chained molecules of glucose molecules create molecules of _glycogen_, cellulose, or starch.
- Cellulose is called a _poly_saccharide.
- Glycogen is called a _poly_saccharide.
- Cellulose and glycogen differ in the way that _glucose_ molecules are bonded together.
- Starch serves as a means for storing glucose in [■ **plants** ❑ animals].
- Glycogen serves as a means for storing glucose in [❑ plants ■ **animals**].
- Lipids are composed of three _fatty acids_ and one _glycerol_ molecule.

QUESTIONS FROM PAGE 32

- At *present*, the gas of highest concentration in the Earth's atmosphere is [❑ helium ❑ oxygen ■ **nitrogen**].
- According to the heterotroph hypothesis, oxygen [❑ was ■ **was not**] a chief component of the atmosphere when life first began.
- According to the heterotroph hypothesis, hydrogen [■ **was** ❑ was not] a chief component of the atmosphere when life first began.
- According to the heterotroph hypothesis, methane [■ **was** ❑ was not] a chief component of the atmosphere when life first began.
- According to the heterotroph hypothesis, ammonia [■ **was** ❑ was not] a chief component of the atmosphere when life first began.
- According to the heterotroph hypothesis, water [■ **was** ❑ was not] a chief component of the atmosphere when life first began.

QUESTIONS FROM PAGE 34

The cell's three areas are the:

- 1. cell _wall_ (in plants and bacteria) and _membrane_ in animal cells.
- 2. _cytoplasm_, and
- 3. nu_cleus_.

In particular,

- The chromosomes are located in the _nucleus_.
- The organelles are located in the _cytoplasm_.
- A plant or bacterial cell has as its outer boundary the cell _wall_ and the cell _membrane_.
- An animal cell has as its outer boundary a cell _membrane_.

QUESTIONS FROM PAGE 36

- As its outermost layers, the [■ **plant** ☐ **animal**] cell has *both* a cell wall and a cell membrane.
- As its outermost layer, the [☐ **plant** ■ **animal**] cell has *only* a cell membrane and no cell wall.
- A cell membrane is made of _protein_ and _lipid_.
- The process of passive diffusion [☐ **does involve** ■ **does not involve**] the expenditure of energy.
- In the process of facilitated transport, a substance crosses the cell membrane with the help of a _carrier_ molecule.
- In the process of facilitated transport, the substance that crosses the cell membrane [☐ **must be** ■ **need not be**] soluble in lipid.
- The process of _active_ _transport_ requires the expenditure of energy.
- In the process of active transport, movement [■ **does** ☐ **does not**] proceed against a concentration gradient.
- In the processes of passive diffusion and facilitated transport, movement [☐ **does** ■ **does not**] proceed against a concentration gradient.

QUESTIONS FROM PAGE 41

- The system of channels and tubes that run throughout the cytoplasm is called, generally, the _endoplasmic_ _reticulum_.
- The _endoplasmic reticulum_ may be designated as _smooth_ or _rough_.
- Within the cytoplasm, ribosomes are the sites of _protein_ _synthesis_.
- Within the cell, ribosomes are located on [☐ **smooth** ■ **rough**] _endoplasmic_ _reticulum_.

QUESTIONS FROM PAGE 43

- Enzymes promote chemical reactions by bringing _reactants_ together in order that they may form products.

- The fact that enzymes interact with reactants by physically fitting together has given rise to the phrase _lock_ and _key_ theory.
- Enzymes are known as organic _catalysts_.
- The physical location at which a substrate attaches to an enzyme is called the enzyme's _active_ _site_.
- When an enzyme has catalyzed a chemical reaction and the products are formed, the *enzyme itself* [☐ **is** ■ **is not**] consumed and is [☐ **unavailable** ■ **available**] to catalyze additional reactions.

QUESTION FROM PAGE 44

- The fact that a given enzyme is designed to catalyze only a particular chemical reaction is described by the phrase _enzyme_ _specificity_.

QUESTIONS FROM PAGE 45

- Glycogen serves as a "bank" that stores _glucose_.
- Glycolysis takes energy from a molecule of _glucose_ and uses it to produce some ATP from _ADP_ and _phosphate_.
- The process of glycolysis produces ATP and converts one molecule of glu_cose_ to two molecules of _pyruvic_ acid.
- The process of glycolysis [☐ **does** ■ **does not**] require oxygen.

QUESTIONS FROM PAGES 46–47

- The principal substance that enters the Krebs cycle is called _acetyl_ _CoA_.
- The Krebs cycle draws energy from _pyruvic_ _acid_ and produces some _ATP_.
- The Krebs cycle [■ **does** ☐ **does not**] require oxygen.
- The Krebs cycle occurs [☐ **before** ■ **after**] glycolysis occurs.
- The Krebs cycle leaves the cell with CO_2, NA_DH_ and FA_DH_$_2$.

QUESTIONS FROM PAGES 48–49

- Electron transport describes a process in which _electrons_ from NA_DH_ and _FADH_$_2$ are "handed down" through a series of _carrier_ molecules.
- Electron transport [■ **does** ☐ **does not**] require oxygen.
- At the completion of electron transport the final recipient of the transported electrons is _oxygen_, which forms _water_.
- The electron transport process takes energy from NADH and $FADH_2$, using it to form _ATP_ through a process called _oxidative_ phosphor_ylation_.

- The electron transport carrier molecules require <u>iron</u>, and an inadequate supply of <u>iron</u> may cause the process to operate improperly.
- The cellular organelle most closely associated with cellular respiration is the <u>mitochondrion</u>.

QUESTIONS FROM PAGE 50

- <u>An</u>aerobic organisms [☐ do ■ do not] conduct the *Krebs cycle*.
- <u>An</u>aerobic organisms [■ do ☐ do not] conduct *glycolysis*.
- In <u>an</u>aerobic organisms the process of glycolysis is followed by a process called <u>fermentation</u>.
- The process of fermentation produces lac<u>tic acid</u> or <u>ethyl alcohol</u> from the <u>pyruvic</u> acid that results from glycolysis.
- If an aerobic organism has an *adequate* oxygen supply it [☐ does ■ does not] conduct fermentation.
- If an aerobic organism has an <u>in</u>adequate oxygen supply it [■ does ☐ does not] conduct fermentation, and pain may result from the accumulation of <u>lactic</u> acid.
- <u>An</u>aerobic respiration produces [■ less ☐ more] ATP than does aerobic respiration.

QUESTIONS FROM PAGE 51

- All chromosomes are made of <u>DNA</u> and <u>protein</u>.
- All chromosomes [☐ are ■ are not] identical.

QUESTIONS FROM PAGE 52

- A chromosome's structure is described with the phrase <u>double helix</u>.
- Watson and Crick are famous for discovering the structure of <u>DNA</u>.
- The two DNA strands found in a chromosome are said to be <u>complements</u> of each other.
- The fact that double-stranded DNA is shaped as a double helix was discovered by <u>Watson</u> and <u>Crick</u>.

QUESTIONS FROM PAGE 52

The four nucleotide bases are
- 1. <u>adenine</u>,
- 2. <u>cytosine</u>,
- 3. <u>guanine</u>, and
- 4. <u>thymine</u>.

QUESTIONS FROM PAGE 55

- Guanine forms a base pair with <u>cytosine</u>.
- Thymine forms a base pair with <u>adenine</u>.
- Cytosine forms a base pair with <u>guanine</u>.
- Adenine forms a base pair with <u>thymine</u>.

QUESTIONS FROM PAGE 55

- If a particular region of a chromosome has, on *one* of its strands, the base sequence adenine-cytosine-guanine, the complementary strand has the base sequence
  (A) guanine-guanine-thymine
  (B) adenine-cytosine-guanine
  (C) thymine-guanine-cytosine
  (D) cytosine-guanine-thymine

QUESTIONS FROM PAGE 57

Designate the order of events through which a DNA molecule replicates by writing the numbers 1, 2, 3, and 4 next to the appropriate statements.

<u>2</u> Each strand serves as a template.

<u>1</u> The double helix "unwinds" and its DNA strands separate.

<u>4</u> The newly formed double-stranded molecule twists to form a double helix.

<u>3</u> Nucleotides align themselves along the template according to the base-pairing rules.

QUESTIONS FROM PAGE 57

- If two individuals are of the same species, then the chromosomes in one individual [☐ are ■ are not] identical to the chromosomes of the other.
- If two cells are taken from the same individual, the chromosomes in one cell [■ are ☐ are not] identical to the chromosomes of the other.

QUESTIONS FROM PAGE 58

- The two members of a chromosome pair are called <u>homologous</u> chromosomes.
- Homologous chromosomes [■ are ☐ are not] similar to each other.
- Homologous chromosomes [☐ are ■ are not] identical.

QUESTIONS FROM PAGE 59

- An organism's genes are contained in its <u>chromosomes</u>.
- The chromosomes exert their influence over the organism by directing the production of [☐ carbohydrates ■ proteins].
- Control over the organism's synthesis of proteins is particularly significant because some proteins serve as <u>enzymes</u> which, in turn, determine the <u>chemical</u> reactions that the cell will conduct.

QUESTIONS FROM PAGE 60

- RNA nucleotides [☐ do ■ do not] contain the exact same bases as do DNA nucleotides.
- <u>R</u>NA [☐ is ■ is not] a double-stranded molecule.
- <u>D</u>NA [■ is ☐ is not] a double-stranded molecule.

QUESTIONS FROM PAGES 61–62

- The creation of an mRNA strand requires first that a strand of <u>DNA</u> separate from its complementary strand.
- mRNA [■ is ☐ is not] formed from a *template* that arises from a <u>D</u>NA molecule.
- mRNA [☐ is ■ is not] formed from a *template* that arises from a <u>R</u>NA molecule.
- When mRNA is formed from a template, DNA adenine bases associate themselves with RNA <u>uracil</u> bases, and DNA cytosine bases associate themselves with RNA <u>guanine</u> bases.
- The process by which a DNA template creates an RNA molecule is called <u>transcription</u>.
- mRNA is a [■ single- ☐ double-] stranded molecule.

QUESTIONS FROM PAGES 63–64

- mRNA leaves the nucleus after it is formed and locates itself on a <u>ribosome</u>, which is the site of protein synthesis.
- A series of three mRNA nucleotides that "codes for" a particular amino acid is called a <u>codon</u>.
- For every amino acid, there [■ is ☐ is not] a codon.

- Once a series of amino acids has aligned itself along a ribosome in accordance with the associated sequence of mRNA codons, <u>peptide</u> bonds form between adjacent amino acids to form a <u>polypeptide</u>.

QUESTIONS FROM PAGE 65

- During a stage called interphase [■ **all** ☐ **only some**] of the cell's chromosomes replicate.
- According to modern biological terminology, a human cell, after interphase, has a total of [☐ **92** ■ **46**] chromosomes, each chromosome having at its center a <u>centromere</u> that joins the chromatids together.
- A human cell, after interphase, has a total of <u>46</u> chromosomes, each made up of two <u>chromatids</u>.
- DNA replication is [☐ **the only** ■ **one of many**] process(es) that take(s) place during interphase.

QUESTIONS FROM PAGE 68

- During a stage called <u>interphase</u>, all of a cell's chromosomes replicate.
- During prophase, the <u>centrioles</u> move away from one another toward opposite sides of the cell.
- The spindle apparatus forms during a stage called [☐ **anaphase** ■ **prophase** ☐ **metaphase**].
- The cell's cytoplasm actually divides during the stage called <u>telophase</u>.
- The centromere divides during a stage called <u>anaphase</u>.
- Cytokinesis involves the division of [☐ **chromosomes** ■ **cytoplasm**].
- The cell's chromosomes become visible during a stage called <u>prophase</u>.
- Duplicate chromosomes separate from one another and move to opposite poles of the cell during a stage called <u>anaphase</u>.

QUESTIONS FROM PAGE 69

- The word [☐ **haploid** ■ **diploid**] refers to a cell for which each chromosome has a homologous chromosome to form a chromosome pair.
- The word [■ **haploid** ☐ **diploid**] refers to a cell for which each chromosome does NOT have a homologous chromosome to form a chromosome pair.
- If, for a particular organism, the diploid number of chromosomes is 10, the haploid number is <u>5</u>.

QUESTIONS FROM PAGE 72

- Prophase of meiosis [■ is ❏ is not] similar to prophase of mitosis.
- The first metaphase of meiosis differs from metaphase of mitosis in that a <u>pair</u> of chromosomes line up on each spindle fiber.
- The first anaphase of meiosis differs from anaphase of mitosis in that centromeres [❏ do ■ do not] divide.
- During the second metaphase, anaphase, and telophase of meiosis, <u>centromeres</u> divide, cells divide, and the resulting four daughter cells contain <u>23</u> chromosomes, each of which [❏ does ■ does not] have a homologous partner.
- The four spermatogonia that result from spermatogenesis are <u>hap</u>loid.

QUESTIONS FROM PAGE 73

- The cell that gives rise to an ovum is located in the <u>ovary</u> and is called a <u>primary oocyte</u>.
- Polar bodies are those cells that [■ disintegrate during oogenesis ❏ remain after other ova have disintegrated].
- Oogenesis ultimately gives rise to [■ one ovum ❏ four ova] with the [■ haploid ❏ diploid] number of chromosomes.
- Gametogenesis includes <u>oogenesis</u> and <u>spermatogenesis</u>, and produces <u>ova</u> and <u>sperm</u>, respectively.

QUESTIONS FROM PAGE 74

- A spermatozoan is a [■ gamete ❏ zygote], and is [■ haploid ❏ diploid].
- An ovum is a [■ gamete ❏ zygote], and is [■ haploid ❏ diploid].
- When a spermatozoan and an ovum merge, they undergo the process of <u>fertilization</u>, and give rise to a [❏ gamete ■ zygote], which is [❏ haploid ■ diploid].

CHAPTER 4: ANSWERS

QUESTIONS FROM PAGE 76

- The term "gene" refers to that portion of a chromosome that ultimately produces one [❏ amino acid ■ protein] via mRNA and the assembly process that occurs on the ribosome.
- A single gene is responsible, ultimately, for the production of a single <u>protein</u>.

- Proteins are most significant for the organism in that they function as <u>enzymes</u>, which catalyze biological reactions.

QUESTIONS FROM PAGE 80

- The term [■ **phenotype** ☐ **genotype**] refers to an organism's observable traits.
- The term "allele" [☐ **is** ■ **is not**] precisely synonymous with the term gene.
- The term <u>allele</u> refers to the fact that the gene responsible for a particular observable trait might exist in more than one version.
- If, in a particular organism, one allele on one member of a homologous chromosome pair codes for blue eye color and a corresponding allele on the other codes for brown eye color, the organism is said to be [☐ **homozygous** ■ **heterozygous**] for eye color.
- If an organism is heterozygous for eye color, with one allele coding for green and the other allele coding for gray, the organism will have [■ **green eyes** ☐ **gray eyes**] if *green* is dominant, and [☐ **green eyes** ■ **gray eyes**] if *gray* is dominant.
- If, for a particular species, the allele that produces a disease called erythemia is dominant and the corresponding allele that produces the absence of disease (a normal organism) is recessive, then:
  (a) an organism with a genotype that is heterozygous for the trait will have the phenotype: [☐ **normal** ■ **erythemia**].
  (b) an organism with a genotype that is homozygous for the dominant allele will have the phenotype: [☐ **normal** ■ **erythemia**].
  (c) an organism with a genotype that is homozygous for the recessive allele will have the phenotype: [■ **normal** ☐ **erythemia**].
  (d) an organism with a phenotype that is normal must have a genotype that is [■ **homozygous** ☐ **heterozygous**].

QUESTIONS FROM PAGE 84

- Parent plant A has phenotype [☐ **red flowers** ■ **yellow flowers**].
- Parent plant B has phenotype [☐ **red flowers** ■ **yellow flowers**].
- Parent plants A and B [■ **can** ☐ **cannot**] possibly produce a plant with red flowers [☐ **because** ■ **even though**] both A and B have [☐ **red flowers** ■ **yellow flowers**].
- The likelihood that parent plants A and B will produce a plant with red flowers is <u>25</u> %.

- The likelihood that parent plants A and B will produce a plant with yellow flowers is _75_ %.

QUESTIONS FROM PAGE 86

- Parent plant A has phenotype [❑ red flowers ■ yellow flowers].
- Parent plant B has phenotype [❑ red flowers ■ yellow flowers].
- Parent plants A and B [❑ can ■ cannot] possibly produce a plant with red flowers.
- The likelihood that parent plants A and B will produce a plant with red flowers is _0_ %.
- The likelihood that parent plants A and B will produce a plant with yellow flowers is _100_ %.

QUESTIONS FROM PAGE 87

- If an organism is born with nonoily feathers, it [❑ is ■ is not] possible that *both* of its parents were homozygous for oily feathers.
- If an organism is born with oily feathers, it [■ is ❑ is not] possible that *both* of its parents were homozygous for oily feathers.
- If an organism is born with oily feathers, it [■ is ❑ is not] possible that *one* of its parents was homozygous for nonoily feathers.
- If an organism is born with nonoily feathers and its parent A had oily feathers, then
  (a) the genotype for parent A was ___Oo___,
  (b) the genotype for parent B was ___Oo___, or ___oo___,
  (c) the phenotype for parent A was ___oily___,
  (d) the phenotype for parent B was ___oily___, or ___nonoily___.

QUESTIONS FROM PAGE 89

- The laws of genetics were drawn up based on the experiments of _Gregor_ _Mendel_.
- Modern understanding of genetics is attributable largely to the work of [❑ Albert Einstein ■ Gregor Mendel ❑ Charles Darwin].

QUESTIONS FROM PAGE 89

- A male person receives from his father a(n) [❑ X ■ Y] chromosome.
- A male person [❑ may ■ must] have the genotype [■ XY, ❑ XY or YY].

- A female person [☐ may ■ must] have the genotype [■ XX, ☐ XX or XY].
- In terms of sex, all persons male or female receive from their mothers a(n) X chromosome.
- In terms of sex, all females receive from their fathers a(n) X chromosome.

QUESTIONS FROM PAGE 92

- With reference to the parents whose genotypes are shown above,
  (a) <u>50</u> % of children are likely be male.
  (b) <u>50</u> % of children are likely to be female.
  (c) There is [☐ no ■ some] likelihood that a child will have phenotype: hemophilia.
  (d) There is [■ no ☐ some] likelihood that a female child will have phenotype: hemophilia.
  (e) There is [☐ no ■ some] likelihood that a male child will have phenotype: hemophilia.
  (f) The likelihood that there will be born a male child with hemophilia is <u>25 %</u>.

CHAPTER 5: ANSWERS

QUESTIONS FROM PAGE 96

- Vertebrates [■ do ☐ do not] have skeletons.
- Vertebrates have [■ endoskeletons ☐ exoskeletons ☐ no skeletons].
- Mammals have [■ endoskeletons ☐ exoskeletons].
- Birds have [■ endoskeletons ☐ exoskeletons].

QUESTIONS FROM PAGES 96–97

- Vertebrates have an [■ endoskeleton ☐ exoskeleton], and the arthropods have an [☐ endoskeleton ■ exoskeleton].
- Arthropods have an [☐ endoskeleton ■ exoskeleton] composed of <u>chitin</u>.
- The body of a reptile derives shape and support from an [■ endoskeleton ☐ exoskeleton], and the body of an insect derives shape and support from an [☐ endoskeleton ■ exoskeleton].

- An endoskeleton is characteristic of _vertebrates_, and an exoskeleton is characteristic of _arthropods_.

QUESTIONS FROM PAGE 99

- Food travels the human digestive tract by passing through these structures, in this order:
    1. _mouth_,
    2. _esophagus_,
    3. _stomach_,
    4. _small intestine_,
    5. _large intestine_,
    6. _rectum_.
- The word "colon" is a synonym for [■ large ☐ small] intestine.
- The phrase "alimentary canal" is synonymous with the phrase "_digestive tract_."

QUESTION FROM PAGE 100

- The stomach is _acidic_, because its walls secrete hydrochloric _acid_.
- The pH of the stomach is approximately [■ 1 ☐ 7 ☐ 14].

QUESTIONS FROM PAGES 101–102

- The salivary glands of the mouth secrete an enzyme called _amylase_, which functions in the digestion of _carbohydrate_.
- The stomach secretes [■ acid ☐ base], and it is therefore [■ acidic ☐ basic].
- The stomach [■ does ☐ does not] secrete a digestive enzyme.
- Pepsin is an _enzyme_, secreted by the _stomach_. It serves in the digestion of _protein_.
- From the stomach chyme enters the _small intestine_.
- Within the small intestine additional enzymes [■ do ☐ do not] appear.
- Enzymes that appear in the small intestine serve in the digestion of _protein_, _carbohydrate_, and _fat_.
- The liver produces _bile_, which is stored in the _gall bladder_.
- Bile is sent to the small intestine from the _gall bladder_, and serves to emulsify _fat_.
- The products of digestion are absorbed into the blood stream through the walls of the _small intestine_.
- The large intestine's primary function is to _reabsorb_ _water_.

QUESTIONS FROM PAGES 104–105

- Photosynthesis is the process by which green plants *produce* [❑ water and carbon dioxide ■ glucose] *from* [■ water and carbon dioxide ❑ glucose].

- Photosynthesis consists of (1) a <u>light</u> reaction in which energy is taken from sunlight in order to form NADPH and ATP and O_2, and (2) a <u>dark</u> reaction in which the energy from ATP and NADPH is used to form <u>glucose</u> from <u>carbon dioxide</u> and water.

- The pigment that enables green plants to "trap" energy from sunlight is called <u>chlorophyll</u>, which within the cell is located in the <u>chloroplasts</u>.

- When the plant cell conducts the photosynthetic dark reaction, it derives energy from the substances <u>NADPH</u> and <u>ATP</u> in order to form glucose from carbon dioxide and water.

- Because green plants can conduct photosynthesis and animals cannot, plants are called [■ autotrophs ❑ heterotrophs], and animals are called [❑ autotrophs ■ heterotrophs].

- Chloroplasts belong to a class of organelles called <u>plastids</u>.

- Stomate size is regulated by [❑ palisade cells ❑ spongy cells ■ guard cells].

- Photosynthesis is carried out primarily in the <u>palisade</u> layer of the leaf.

- Gas exchange is carried out primarily in the <u>spongy</u> layer of the leaf.

QUESTIONS FROM PAGES 106–107

- [❑ Red blood cells ■ White blood cells] function in the immune system.

- [■ Red blood cells ❑ White blood cells] contain hemoglobin.

- Hemoglobin delivers <u>oxygen</u> throughout the body.

- Hemoglobin contains <u>iron</u>.

- White blood cells manufacture antibodies.

- Antibodies serve to fight <u>infec</u>tion by <u>bact</u>eria and <u>viruses</u> and other microorganisms.

- T-lymphocytes are [■ white ❑ red] blood cells.

- AIDS patients have inefficient [❑ red blood cells ■ white blood cells] and therefore have a defective <u>immune</u> system.

QUESTIONS FROM PAGES 108–109

- Respiration involves the [■ inhalation ❑ exhalation] of oxygen and the [❑ inhalation ■ exhalation] of carbon dioxide and water.

- Within the lung, oxygen is inhaled and then crosses the wall of the al<u>veolus</u> to enter the blood in a pulmonary <u>capillary</u>.

- Within the lung, <u>carbon</u> <u>dioxide</u> and water cross the wall of a pulmonary capillary to enter the <u>alveolus</u> and then to be exhaled.

- The carbon dioxide and water that reach the lung via the blood stream are produced in the body's cells as a result of cellular <u>respiration</u>.

QUESTIONS FROM PAGE 112

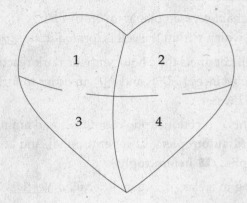

- As shown in the diagram above, the heart's four chambers are the
  1. <u>right</u> <u>atrium</u>,
  2. <u>left</u> <u>atrium</u>,
  3. <u>right</u> <u>ventricle</u>, and
  4. <u>left</u> <u>ventricle</u>.

- Blood leaves the heart at the [■ **left ventricle** ❑ **right atrium**], and enters the heart at the [❑ **left ventricle** ■ **right atrium**].

- When blood exits the heart, it enters a large artery called the <u>aorta</u>, which divides repeatedly to form smaller arteries, arterioles, and finally <u>capillaries</u>.

- The vessels that come into intimate contact with the body's cells in order to conduct exchange of gas and other substances are the [❑ **venules** ■ **capillaries**].

- Capillaries come together to form ven<u>ules</u> and veins, ultimately forming two large veins called the <u>anterior</u> <u>vena</u> <u>cava</u> and the <u>posterior</u> <u>vena</u> <u>cava</u>.

- Blood is conducted back to the heart ultimately by the [■ **anterior and posterior vena cavae** ❑ **aorta**].

QUESTIONS FROM PAGE 115

- Blood that enters the right atrium after touring the entire body is [☐ oxygen-rich ■ oxygen-poor].
- From the right atrium, blood is passed immediately to the <u>right</u> <u>ventricle</u>.
- From the right ventricle [☐ oxygen-rich ■ oxygen-poor] blood is passed to the [■ pulmonary arteries ☐ pulmonary veins], and then to the lungs.
- From the lungs, [■ oxygen-rich ☐ oxygen-poor] blood is passed to the [☐ pulmonary arteries ■ pulmonary veins], and then to the heart's <u>left atrium</u>.
- The pulmonary arteries carry [☐ oxygen-rich ■ oxygen-poor] blood [■ away from ☐ toward] the heart and the pulmonary veins carry [■ oxygen-rich ☐ oxygen-poor] blood [☐ away from ■ toward] the heart.
- The aorta carries [■ oxygen-rich ☐ oxygen-poor] blood [■ away from ☐ toward] the heart and the two vena cavae carry [☐ oxygen-rich ■ oxygen-poor] blood [☐ away from ■ toward] the heart.

QUESTIONS FROM PAGE 116

- [■ Xylem ☐ Phloem] serves primarily to move water up a plant.
- [☐ Xylem ■ Phloem] serves primarily to move carbohydrate and protein up and down a plant.
- Food moves through the <u>phloem</u> in a plant.
- Water moves through the <u>xylem</u> in a plant.
- Vessels located within the [☐ xylem ■ phloem] move protein and carbohydrate in the [☐ upward direction only ☐ downward direction only ■ upward and downward directions].
- Vessels that directly conduct organic compounds in a plant are called [☐ tracheids ■ sieve tube cells].

QUESTIONS FROM PAGE 116

- A plant's roots serve to absorb <u>water</u> and <u>minerals</u>. They also serve as an <u>anchor</u> that maintains the plant's position.

QUESTIONS FROM PAGES 118–119

- A nerve cell is also called a <u>neuron</u>.
- Within a neuron, the mitochondria, ribosomes, and other organelles are located in the [☐ dendrites ■ cell body ☐ axon].

- A neuron conducts a nerve impulse in the direction [□ axon→cell body→dendrite ■ dendrite→cell body→axon].

- Within a neuron the dendrites carry the nerve impulse [■ toward □ away from] the cell body, and the axon carries the nerve impulse [□ toward ■ away from] the cell body.

- If an impulse moves from neuron A to neuron B, then it moves from the [□ dendrites ■ axon] of neuron A to the [■ dendrites □ axon] of neuron B.

- A neuron that transmits an impulse from the central nervous system directly to a muscle is called a(n) <u>motor neuron</u>.

- When an impulse moves from one neuron to the next, the neurons [□ are ■ are not] in physical contact.

- When an impulse moves from one neuron to the next, it is carried by a chemical "<u>messenger</u>."

- One of the substances responsible for moving a nerve impulse from one neuron to the next is called acetyl<u>choline</u>.

- A neuron that is located between a sensory neuron and a motor neuron is called a(n) <u>interneuron</u>.

Questions from page 120

- A resting neuron is relatively positive <u>out</u>side and negative <u>in</u>side.

- When a neuron experiences an action potential its permeability to <u>sodium</u> increases, which results in <u>depolarization</u>.

- A neuron that is depolarized, due to an influx of <u>sodium</u> ions, is relatively <u>negat</u>ive on the outside and <u>posit</u>ive on the inside.

Questions from page 122

- When a portion of a neuron undergoes depolarization, it then undergoes <u>repolariz</u>ation almost immediately.

- <u>Depolarization</u> results from an [■ influx □ efflux] of [■ sodium □ potassium] ions.

- <u>Repolarization</u> involves the [□ influx ■ efflux] of [□ sodium ■ potassium] ions.

- Once a cell has undergone depolarization, *and then* repolarization, it is relatively [■ negative □ positive] on the inside and [□ negative ■ positive] on the outside.

- In terms of relative charge inside and outside, a cell that has undergone <u>repolarization</u> is [■ like □ unlike] a resting neuron.

- A nerve impulse is conducted along an entire neuron because the _action potential_ spreads from dendrite to cell body to axon.

- An action potential spreads because when one portion of a neuron's cell membrane repolarizes, a small portion adjacent to it _de_polarizes and thus experiences an increase in permeability to _sodium_ ions.

QUESTIONS FROM PAGE 124

- The phrase "_central_ nervous system" refers to neurons located *within* the brain and spinal cord.

- The medulla is responsible for coordinating [❏ **thought processes** ■ **breathing rate**].

- The phrase "_peripheral_ nervous system" refers to neurons located *outside* of the brain and spinal cord.

- The cerebellum helps an organism to maintain its [❏ **heart rate** ■ **balance**]

- Neurons of the peripheral nervous system are [❏ **entirely separate from** ■ **connected to**] neurons of the central nervous system.

- Thought processes are carried out by the [❏ **medulla** ❏ **cerebellum** ■ **cerebrum**]

QUESTIONS FROM PAGE 125–126

- If a stimulus reaches a neuron, it [❏ **will** ■ **will not**] necessarily produce an action potential.

- Only a stimulus that reaches the neuron's _threshold_ will cause the neuron to experience an action potential.

- If a neuron experiences a stimulus that fails to reach its threshold, it [❏ **will** ■ **will not**] experience an action potential.

- For some short period of time after a neuron experiences an action potential, it undergoes a _refractory_ period which means it cannot, for that period, experience an additional action potential no matter how strong an impulse it may receive.

- During a neuron's refractory period the neuron [❏ **can** ■ **cannot**] experience an action potential.

- The sodium-potassium pump helps the neuron maintain a high concentration of _potass_ium _in_ternally and a high concentration of _sodium_ _ex_ternally.

- Operation of the sodium-potassium pump [■ **does** ❏ **does not**] require energy.

- [❏ **All** ■ **Some**] neurons are wrapped in a myelin sheath.

- A myelin sheath [■ **speeds** ❏ **slows**] the neuron's conduction of an impulse.

- A myelin sheath is actually composed of _Schwann_ cells.

Questions from pages 130–131

- On each blank line place the number that designates the appropriate hormone.

    A. The islets of Langerhans secrete _5_ and _13_.

    B. The adrenal cortex secretes _7_ and _10_.

    C. The anterior pituitary secretes _4_, _9_, and _2_.

    D. The thyroid gland secretes _6_.

    E. The parathyroid gland secretes _1_.

    F. The posterior pituitary gland secretes _11_ and _12_.

    G. The adrenal medulla secretes _3_ and _8_.

    1. parathyroid hormone
    2. thyroid-stimulating hormone
    3. epinephrine
    4. growth hormone
    5. insulin
    6. thyroid hormone
    7. mineralocorticoids
    8. norepinephrine
    9. ACTH
    10. glucocorticoids
    11. oxytocin
    12. vasopressin
    13. glucagon

- On each blank line place the number that designates the appropriate hormone.

    Hormone

    parathyroid hormone _3_

    thyroid-stimulating hormone _7_
    growth hormone _9_

    insulin _5_
    thyroid hormone _1_

    mineralocorticoids _8_

    ACTH _10_

    Effect

    1. increases body's metabolic rate
    2. increases blood glucose concentration
    3. increases blood calcium concentration
    4. increases reabsorption of sodium and water in the kidney

glucocorticoids _2_

5. decreases blood glucose concentration

oxytocin _6_

6. causes uterus to contract and mammary glands to release milk

vasopressin _8_

7. stimulates thyroid gland to produce and release thyroxine

glucagon _2_

8. causes kidney to retain water

epinephrine _11_

9. targets all tissues and causes tissue growth

norepinephrine _11_

10. stimulates the adrenal cortex

11. prepares body for "fight-or-flight" response

QUESTIONS FROM PAGE 133

- The female menstrual cycle is mediated, chiefly, by secretions from the _an_terior pit_uitary_ gland and the _ova_ry.
- The anterior pituitary [❏ does ■ does not] secrete estrogen.
- The anterior pituitary [■ does ❏ does not] secrete luteinizing hormone.
- A follicle consists of an _oocyte_ and surrounding layers of _cells_.
- The ovary/growing follicle [■ does ❏ does not] secrete estrogen.

Fill in the blanks and boxes appropriately.

- At the end of the follicular phase of the female menstrual cycle, the _anterior pituitary_ gland suddenly increases its production of luteinizing hormone.
- The follicular phase of the female menstrual cycle [❏ begins ■ ends] with a surge in the secretion of luteinizing hormone (LH).
- The luteal surge produces _ovulation_.
- The process by which a mature follicle ruptures and releases its ovum is called _ovulation_.

- One follicle matures during a period called the
  [☐ **luteal surge** ■ **follicular phase**].
- During the follicular phase, the endometrical lining of the uterus
  [■ **builds up** ☐ **sloughs off**].

QUESTIONS FROM PAGE 134

Fill in the blanks and boxes appropriately.

- During the second phase of the female menstrual cycle, the corpus luteum secretes _estrogen_ and _progesterone_. The progesterone gets the uterus ready for pregnancy.
- During the luteal phase of the female menstrual cycle, the ovum, traveling in the fallopian tube [☐ **does** ■ **does not**] secrete estrogen and progesterone.
- During the luteal phase of the female menstrual cycle, the corpus luteum, situated in the ovary, [■ **does** ☐ **does not**] secrete estrogen and progesterone.

QUESTIONS FROM PAGE 134–135

Draw lines connecting the event on the left with the appropriate order of occurrence on the right.

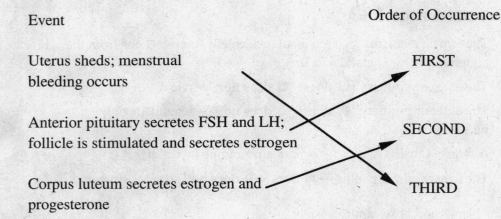

- When the corpus luteum stops secreting estrogen and progesterone, the female experiences her menstrual _period_, which is really a period of _bleeding_ from the uterus.
- The menstrual period (flow phase) of the female menstrual cycle [☐ **is** ■ **is not**] initiated when the anterior pituitary stops secreting luteinizing hormone.
- The menstrual period (flow phase) of the female menstrual cycle [■ **is** ☐ **is not**] initiated when the corpus luteum stops secreting estrogen and progesterone.

- When fertilization and implantation occur, the corpus luteum [■ does ❏ does not] continue to secrete estrogen and progesterone, and the uterine lining [❏ is ■ is not] shed.

QUESTIONS FROM PAGE 135

- The principle male sex hormone is called <u>testosterone</u> and it's secreted by the <u>testes</u>.
- Testosterone promotes the development of the male sex organs, muscles, and <u>secondary sex characteristics</u>, such as facial hair.

QUESTIONS FROM PAGES 136–137

- The ovum is produced by the [❏ uterus ■ ovary].
- Implantation in the uterus occurs [❏ before ■ after] fertilization.
- Fertilization takes place in the [❏ ovary ■ fallopian tube ❏ uterus].
- The gametes arise [■ before ❏ after] the zygote arises.
- When the sperm cell fertilizes the ovum, it produces a <u>zygote</u>.
- The embryo arises [❏ before ■ after] implantation in the uterus.

QUESTIONS FROM PAGE 137

- Correctly order the following sequence of events and structures associated with human embryology.
    A. fertilization <u>3</u>
    B. embryo <u>7</u>
    C. gametes <u>1</u>
    D. ovulation <u>2</u>
    E. implantation <u>6</u> in uterus
    F. zygote <u>4</u>
    G. birth <u>8</u>
    H. cleavage <u>5</u>

QUESTIONS FROM PAGES 137–138

- A developmental structure that follows cleavage, but precedes the gastrula is the [❏ zygote ❏ embryo ■ blastula].
- A developmental stage marked by a series of rapid mitotic divisions is known as <u>cleavage</u>.
- A sphere of cells one layer thick and filled with fluid is called the [■ blastula ❏ gastrula ❏ embryo].
- A developmental stage known as gastrulation produces a structure that [❏ is perfectly spherical ■ bears a depression on one side].
- The correct sequence of a developing embryo is zygote, <u>cleavage</u>, <u>blastula</u>, <u>gastrula</u>, and embryo.

QUESTIONS FROM PAGE 138–139

- The three embryonic germ layers are called <u>ecto</u>derm, <u>meso</u>derm, and <u>endo</u>derm.
- The human eye develops from the germ layer known as [☐ mesoderm ■ ectoderm].
- Human blood vessels develop from the germ layer known as [■ mesoderm ☐ ectoderm].
- The *outer* lining of the human intestine develops from the germ layer known as [■ mesoderm ☐ endoderm].
- The *inner* lining of the human trachea develops from the germ layer known as [☐ ectoderm ■ endoderm].
- Ectoderm gives rise to the human <u>eye</u>, <u>nervous</u> system, and the [■ outer layer ☐ inner layers] of the skin, known as the <u>epi</u>dermis.
- Endoderm gives rise to the inner linings of the human <u>respiratory</u> and <u>diges</u>tive structures.

QUESTIONS FROM PAGE 140

- In the flowering plant, pollen is *produced by and located on* the [☐ stigma ■ anther].
- In the flowering plant, pollen *germinates* at the [■ stigma ☐ anther].
- In order to reach the ovary in a flowering plant, microspores must pass through the [☐ anther ■ pollen tube].
- The female reproductive parts of a flower are the <u>stigma</u>, <u>style</u> and ovary.
- In the flowering plant microspores are housed within a substance called <u>pollen</u>.
- Following fertilization, the ovary develops into the <u>fruit</u>.
- In the flowering plant megaspores are housed within a structure called the <u>ovule</u>.
- In the flowering plant, reproduction requires fertilization of <u>mega</u> spores by <u>micro</u> spores, and results in the formation of a <u>seed</u>.
- The male reproductive parts of a flower are the <u>filament</u>, an<u>ther</u> and <u>pollen</u>.

QUESTIONS FROM PAGE 141

- When an animal makes a "decision" early in life based on some critical cue, it has undergone a learning process called <u>imprinting</u>.
- If by repeated trial a cat comes to know that if it rings the bell tied to to its owner's front door, the owner will open the door, it [☐ has ■ has not] learned by imprinting.

QUESTIONS FROM PAGE 142

- If a dog is scolded each time it steps on a living room carpet and for that reason the dog stops stepping on the carpet, then the dog [■ **has** ❑ has not] experienced conditioning.

- If a rat is rewarded each time it presses a particular bar and for that reason presses the bar repeatedly, the rat [■ **has** ❑ has not] experienced conditioning.

QUESTIONS FROM PAGE 142

- Consider a dog tied to a tree. In its travels it circles the tree clockwise several times and finds that its leash has been shortened because it is wrapped several times around the tree's trunk. The animal has never faced the situation before, but determines to walk counterclockwise around the tree several times in order to restore its leash to its original length. The animal has demonstrated [❑ **imprinting** ❑ **conditioning** ■ **reasoning/insight**].

- After each description of learning identified below, write:
  (**A**) if it represents imprinting.
  (**B**) if it represents conditioning.
  (**C**) if it represents insight.

1. A dog learns that if it brings the newspaper into the house each evening it gets a bone. _B_

2. A man wishes to turn a screw and, having no screwdriver, realizes that he can use the edge of a dull knife as a substitute. _C_

3. A bird is born and sees, in its nest, other birds. The bird treats itself, from then on, as belonging to the same species as these other birds. _A_

4. A cockroach learns to run from light because every time a light comes on someone tries to step on it. _B_

QUESTIONS FROM PAGE 143–144

- If the stem of a plant grows toward a light source it exhibits the phenomenon of [❑ **negative geotropism** ■ **phototropism**], which is caused by a class of hormones called _auxins_.

- If the root of a plant grows toward a water source it exhibits the phenomenon of [■ **hydrotropism** ❑ **negative geotropism** ❑ **positive geotropism**], which is caused by a class of hormones called _auxins_.

- If a tree should fall but remain alive and start turning its growth upward toward the sky, it exhibits the phenomenon of [❑ **hydrotropism** ■ **negative geotropism** ❑ **positive geotropism**], which is caused by a class of hormones called _auxins_.

- If a tree should fall but remain alive and its roots should "insist" on growing straight downward, it exhibits the phenomenon of
[❏ hydrotropism ❏ negative geotropism ■ positive geotropism], which is caused by a class of hormones called auxins.

QUESTIONS FROM PAGES 144–145

- The fact that plants (or animals) have an innate sense of time is attributable to a "biological clock."
- When a biological clock dictates behavior conforming to a *24-hour day*, the clock is said to reflect a circadian rhythm.
- Certain mammals predictably hibernate for certain months of the year and do not hibernate in others. Such behavior [❏ does ■ does not] reflect a circadian rhythm.
- Certain birds remain in their nests throughout the day and emerge only at night. Such behavior [■ does ❏ does not] reflect a circadian rhythm.

QUESTIONS FROM PAGE 145

- An animal that releases a chemical which in turn affects the behavior of another animal of its species is said to release a pheromone.
- If an animal produces and secretes into its own blood stream some chemical that produces an effect on *its own* behavior, that chemical [❏ is ■ is not] a pheromone.
- If within a species, one individual produces and secretes a chemical that produces an effect on another that chemical [■ is ❏ is not] a pheromone.

QUESTIONS FROM PAGE 146

- Whether they harm or help one another, two organisms of different species, living in close association, are said to be in a symbiotic relationship.
- When two organisms live in intimate association and each confers a benefit on the other, the two organisms are in a symbiotic relationship that is mutualistic.
- The term "symbiosis" [■ does ❏ does not] describe any arrangement in which two organisms of different species live in close association.
- Mutualism [■ is ❏ is not] a form of symbiotic relationship.
- Commensalism [■ is ❏ is not] a form of symbiosis.
- If two organisms live in intimate association and one gains a benefit while the other suffers harm, the two organisms [❏ do ■ do not] exhibit commensalism.

- If two organisms live in intimate association and one gains a benefit while the other suffers harm, the two organisms [■ do ❑ do not] exhibit parasitism.
- Parasitism [■ is ❑ is not] a form of symbiotic relationship.

QUESTIONS FROM PAGE 147

- [■ Some ❑ All] protists can conduct photosynthesis.
- [■ Some ❑ All] protists resemble animal cells.
- [❑ Some ■ All] protists are single-celled.
- Fish [❑ are ■ are not] protists.
- The euglena [■ is ❑ is not] a protist.
- The am<u>oeba</u> and the para<u>mecium</u> are protists.
- The am<u>oeba</u>, para<u>mecium</u>, and eu<u>glena</u> are [❑ multicellular ■ unicellular].

QUESTIONS FROM PAGE 147

- A fungus has [❑ one ■ many] nuclei.
- Most fungi reproduce by forming spores, but yeasts reproduce by <u>budding</u>.

QUESTIONS FROM PAGE 148

- Bacteria are <u>pro</u>karyotic cells.
- Bacterial cells [❑ do ■ do not] have mitochondria.
- Bacterial cells [❑ do ■ do not] have a Golgi apparatus.
- Bacterial cells [■ do ❑ do not] have a cell wall.

QUESTIONS FROM PAGE 149

Conjugation: #_3_

Transduction: #_1_

Transformation: #_2_

1. Virus carries DNA from one bacterium to another
2. One bacterium transfers genetic material to another, causing a change in pheno-type in the second bacterium
3. One bacterium donates a fragment of DNA to another

QUESTIONS FROM PAGE 149

- [■ No bacteria   ☐ Some bacteria] are carnivores.
- [■ No bacteria   ☐ Some bacteria] are herbivores.
- [☐ No bacteria   ■ Some bacteria] are autotrophs.
- [☐ Few bacteria ■ Most bacteria] are heterotrophs.
- [☐ No bacteria   ■ Some bacteria] are saprophytes.
- [☐ No bacteria   ■ Some bacteria] are parasites.

QUESTIONS FROM PAGE 149

Obligate anaerobe #_2_

Facultative anaerobe #_3_

Obligate aerobe #_1_

1. derives energy via aerobic respiration only; cannot survive without oxygen
2. derives energy via fermentation only; cannot survive in the presence of oxygen
3. can derive energy via aerobic respiration or fermentation; can survive with or without oxygen

QUESTIONS FROM PAGES 150–151

- A viral coat is made of [■ protein only   ☐ lipid and protein].
- [■ TRUE   ☐ FALSE] All viruses contain DNA or RNA.
- [☐ Some viruses   ■ All viruses] contain nucleic acid.
- Viruses [☐ do   ■ do not] have, within them, all of the cellular machinery necessary for their own reproduction.
- [■ TRUE   ☐ FALSE] In order to reproduce themselves, viruses make use of cellular machinery belonging to a host cell.
- In reproducing itself a virus injects its _nucleic_ acid into a host cell.
- Which of the following correctly orders the events associated with viral reproduction?

    (A)
    1. Viral nucleic acid is replicated using host cell machinery
    2. New viral particles lyse host cell and emerge
    3. Protein coat attaches to host cell
    4. Virus injects nucleic acid into host cell
    5. New protein coats are formed

(B) 1. Protein coat attaches to host cell
2. Virus injects nucleic acid into host cell
3. Viral nucleic acid is replicated using host cell machinery
4. New protein coats are formed
5. New viral particles lyse host cell and emerge

(C) 1. New protein coats are formed
2. Virus injects nucleic acid into host cell
3. Viral nucleic acid is replicated using host cell machinery
4. New viral particles lyse host cell and emerge
5. Protein coat attaches to host cell

(D) 1. Protein coat attaches to host cell
2. Virus injects nucleic acid into host cell
3. New viral particles lyse host cell and emerge
4. Viral nucleic acid is replicated using host cell machinery
5. New protein coats are formed

QUESTIONS FROM PAGE 152

- The phrase "genetic _variability_" refers to the fact that for any gene pertaining to any population of any species, the gene pool features a variety of alleles.

- The phrase "_genetic_ variability" refers to the fact that within any population of any species, genotypes vary.

- Genetic variability [❏ is ■ is not] caused by a species' tendency to improve its relationship with its environment.

- Genetic variability [❏ is ■ is not] caused by a species' inability to adapt to existing environmental conditions.

- Genetic variability [■ is ❏ is not] attributable to random mutation.

- Genetic variability [■ is ❏ is not] a property of all populations.

QUESTIONS FROM PAGE 153–154

- Evolution means change in a population's _gene_ _pool_.
- Evolution [❏ always ■ sometimes] results in the production of a new species.

- If a population is geographically divided, it [□ cannot ■ may] give rise to two separate species.
- Speciation [■ is □ is not] a product of evolution.
- Speciation [■ increases □ decreases] biological diversity.
- Divergent evolution can result from [□ only physical □ only behavioral ■ both physical and behavioral] selection pressures from the environment.
- In the course of divergent evolution, closely related species become [□ more ■ less] similar to each other with regard to behaviors and traits.

QUESTIONS FROM PAGE 156

- The science of classification is called [□ taxidermy ■ taxonomy].
- The conventional ordering of phylogeny is kingdom, phylum, _class_, _order_, _family_, _genus_, _species_.
- The binomial system of nomenclature was conceived by [□ Charles Darwin ■ Carolus Linnaeus].
- The members of a kingdom [□ do ■ do not] have more in common than do the members of an order.
- The members of an order [■ do □ do not] have more in common than do the members of a class.

QUESTIONS FROM PAGE 159

- With respect to the evolution of life forms, reptiles appeared [□ before ■ after] amphibians.
- With respect to the evolution of life forms, fish appeared [■ before □ after] birds, and birds appeared [■ before □ after] mammals.
- The order in which life forms appeared on Earth is (1) fish, (2) _amphibians_, (3) _reptiles_, (4) _birds_, (5) _mammals_.

QUESTIONS FROM PAGE 159

- The term "niche" [□ does ■ does not] refer solely to the physical habitat in which an organism lives.
- The term "niche" [■ does □ does not] refer, abstractly, to all aspects of an organism's relationship to the biological system in which it lives.
- The term "niche" is synonymous with [■ habitat □ food chain □ lifestyle].

Questions from pages 161

- In a food chain consisting of grass, mice, snakes, and haeks, the mice are <u>primary consumers</u> and the hawks are tertiary <u>consumers</u>.

- In a food chain consisting of grass, grasshoppers, frogs, and bass, grasshoppers are [■ **more** ❑ **less**] numerous than frogs, and frogs are [❑ **more** ■**less**] numerous than bass.

- "Heterotroph" is synonymous with <u>consumer</u> and "Autotroph" is synonymous with <u>producer</u>.

- The energy available to an organism [❑ **increases** ■ **decreases**] with each successive feeding level of a food chain, beginning with producers.

- Decomposers include <u>bacteria</u> and <u>fungi</u>.

- Primary consumers are called <u>herbivores</u> and secondary consumers are called <u>carnivores</u>.

- An herbivore is a(n) [❑ **autotroph** ■ **heterotroph**].

Questions from page 162–163

- The phrase "ecological succession" refers to the development or alteration of biological <u>communities</u>.

- The organism that first appears as "founder" of a biological community is called a <u>pioneer</u> organism.

- When a plant community begins on a barren rock, the pioneer organism is most often [■ **lichen** ❑ **moss**].

- Each plant community that inhabits an area contributes to the process of ecological succession by modifying the environment, making it [■ **more** ❑ **less**] favorable for itself, and [■ **more** ❑ **less**] favorable for another plant community.

- In the ecological succession depicted below, the climax community is composed of <u>deciduous trees</u>.

    pond → cattails → moss and sedges → shrubs → pine trees → deciduous trees

- In an ecological succession, each new plant community in an area [❑ **coexists with** ■ **replaces**] the previous plant community.

- In the ecological succession depicted below, impermanent plant communities include <u>pioneer organism</u>, and <u>plant communities A, B, and C</u>.

Questions from pages 163–164

- The biome with frozen soil year-round and the fewest number of trees is the <u>tundra</u>.

- The biome with the least rainfall and a population of succulent plants is the _desert_.

- The biome with the greatest number of organisms is the _tropical rainforest_.

- The biome characterized by a large number of conifers (evergreen trees) is the _taiga_.

- The tropical rainforest has [■ **more** ❏ **fewer**] life forms than the tundra.

- The taiga [■ **is** ❏ **is not**] rich in conifers.

- Deer, black bear, and raccoon are found in the [❏ **tundra** ❏ **rainforest** ■ **temperate deciduous forest**], while caribou, beaver, and moose are found in the [❏ **tundra** ■ **taiga** ❏ **temperate deciduous forest**].

- The primary plant forms of the [❏ **taiga** ■ **tundra** ❏ **desert**] are mosses, lichens, and wildflowers.

- The primary flora of the desert are the succulents, such as _cacti_ and _aloe vera_.

ABOUT THE AUTHOR

Theodore Silver holds a medical degree from the Yale University School of Medicine, a bachelor's degree from Yale University, and a law degree from the University of Connecticut.

Dr. Silver has been intensely involved in the field of education, testing, and test preparation since 1976 and has written several books and computer tutorials pertaining to those fields. He became affiliated with The Princeton Review in 1988 and is chief author and architect of The Princeton Review MCAT preparatory course.

Dr. Silver is Associate Professor of Law at Touro College Jacob D. Fuchsberg Law Center where he teaches the law of medical practice and malpractice, contracts, and federal income taxation.

Diagnostic Test Form

NOTES

NOTES

NOTES

NOTES

NOTES

NOTES

NOTES

NOTES

NOTES

NOTES

Maximize Your Chance Of Playing College Athletics...
Enroll In The Student Athlete Information Link!!

The Student Athlete Information Link (SAIL) is the nation's premier athletic recruiting and information service for student athletes of all abilities. SAIL will help college coaches find you and make sure you are ready when they call!

How does SAIL help maximize your chances of being recruited? Complete the Student Athlete Profile and your credentials become available to every college coach nationwide at no cost to the college. Using the technology of the Internet, you gain national exposure for a very low cost. Your personal webpage includes your academic and athletic accomplishments along with your photo. You may also include a letter of recommendation, newspaper clippings, or other key information.

SAIL is much more than a recruiting service. It is a total information source for college-bound student athletes. The SAILMAIL newsletter provides valuable guidance to you and your parents on topics such as NCAA regulations and academic requirements, and understanding the scholarship and recruiting processes. The SAILMAIL newsletter is absolutely free with all paid SAIL enrollments.

Many college coaches maintain information on potential recruits as early as their sophomore season. Individuals may enroll in SAIL as soon as they complete their first high school season or similar time period for those athletes not playing for their high school. Not only will you increase your exposure by enrolling early, but you will have the benefit of receiving SAILMAIL for the entire time.

WHAT YOU RECEIVE
- Your own personal webpage including your photo
- Exposure to college coaches nationwide
- The SAILMAIL newsletter
- Up to 3 free updates

GET STARTED NOW

Complete the Student Athlete Profile on the back of this page or register on-line at **www.studentathletelink.com** and SAIL will build your personal webpage. You may also include a photograph with your profile at no additional cost—preferably an action photo or one of you in uniform. Finally, you may update your profile three times during the course of the year for free. Upon receiving your profile we will send you the appropriate sport statistics form.

For those interested in enhancing their profile, you may add a letter of recommendation, newspaper clipping, or other such item to your profile. A page of this type can provide coaches with many additional qualities you possess. The fee is only $15 per customized page (8 1/2 x 11).

Return your completed profile, photo, and any customized pages along with your check to SAIL: P.O. Box 2382; Columbia MD 21045. It will take approximately 3-4 weeks for your profile to be placed on the Internet. You may view it by accessing the "Athlete Interact" page of our website.

If you have any questions or would like more information about SAIL, contact us at (410) 418-5380 or visit our website at www.studentathletelink.com.

STUDENT ATHLETE INFORMATION LINK (SAIL)
STUDENT ATHLETE PROFILE

INSTRUCTIONS: Please print clearly. Complete as much information as possible. The profile does not have to be complete to be valuable to college coaches. If you have any questions please contact SAIL at (410) 418-5380 or visit our website at www.studentathletelink.com. For your information, your Social Security Number and address cannot be viewed by the college coach or Internet user. We will send a sport statistic form for you to complete upon receipt of your profile.

PERSONAL INFORMATION

_____ _____ _____ ___-__-____
LAST NAME   FIRST NAME   MI   SOCIAL SECURITY #

_____ _____ _____ _____
ADDRESS   CITY   STATE ZIP CODE

___/___/___   M___ F___   (___)____-_____   Do you want your home phone on the Internet? Y____ N____
BIRTH DATE   GENDER   HOME PHONE   If NO the college coaches' point of contact will be your high school.

___'___" _____LBS. R____ L____   RACE   Caucasian ____   African-Amer. ____   Asian ____
HEIGHT   WEIGHT   DOMINANT HAND (Optional) Hispanic ____   Amer. Indian ____   Other ____

PARENTS ANNUAL INCOME (Optional)   Less than 20,000____   20,000–40,000____   40,000–80,000____   Over 80,000____

E-MAIL ADDRESS _____

ACADEMIC AND ATHLETIC INFORMATION

_____   _____
HIGH SCHOOL   (PRINT FULL HIGH SCHOOL NAME)   ZIP CODE OF SCHOOL

_____   _____   _____ (based on 4.0 scale)   I have applied for NCAA Clearinghouse. (circle one) Yes or No
SAT   ACT   GPA

(____) _____-_____
ATHLETIC DIRECTOR PHONE NUMBER

H. S. GRAD. YR _____   # OF YRS PLAYING VARSITY SPORTS____   # OF VAR. SPORTS PLAYED____   AAU PARTICIPANT Y___ N___

_____   _____   (____) _____-_____
NAME OF CLUB TEAM OR CAMP ATTENDED   NAME OF CLUB COACH   CLUB COACH PHONE NUMBER
(OTHER THAN HIGH SCHOOL)

VERT. JUMP_____in.   SQUAT____lbs.   BENCH____lbs.   SPEED IN 40YDS____
AWARDS: ALL-CONFERENCE/DISTRICT____   ALL-STATE____   ALL-AMERICAN____

Please list the sport(s) you would like to play in college. We will send you the appropriate form so you can provide your statistics to include with your profile. You may include up to three sports.

1._____   2._____   3._____

I understand that the information provided above may be relied on by coaches and/or athletic directors and do hereby verify and attest that the information provided above is accurate to the best of my knowledge.

STUDENT SIGNATURE _____   DATE _____
PARENT'S SIGNATURE _____   DATE _____

PAYMENT INFORMATION: CHECK ENCLOSED ____   VISA/MASTERCARD ____   DISCOVER CARD ____

NEW STUDENT ATHLETE PROFILE FORM @ $30 $____
_____
CREDIT CARD NUMBER   EXP. DATE

CUSTOMIZED PAGE(S) @ $15 PER PAGE (8_ X 11) $____
_____
CREDIT CARD NUMBER   EXP. DATE

TOTAL AMOUNT INCLUDED WITH PROFILE $____
_____
ADDRESS (If different from above)   CITY   STATE   ZIP

Send your profile, photo and customized page(s), along with your check for $30 to SAIL: P.O. 2382; Columbia, MD 21045.
Allow 3-4 weeks for your profile to be processed. Thank you.

www.review.com

Expert Advice

Talk About It

Pop Surveys

www.review.com

Paying for it

www.review.com

www.review.com

THE PRINCETON REVIEW

Getting in

Word du Jour

www.review.com

Find-O-Rama School & Career Search

www.review.com

Best Schools

Finding it

www.review.com

You're a student. Milk it!

The Student Advantage® Member ID - The Official Student Discount Program of The Princeton Review.

Carry it. Show it. Save cash.

Think of it as the National Student ID that gets you huge discounts on the stuff you need and want. Flash it and save all year long at over 15,000 sponsor locations in your area and nationwide. Below are just a few*.

For a complete listing of our sponsors, check out www.studentadvantage.com

To join Student Advantage for just $20 call 1-800-333-2920.

*Benefits are provided by independent suppliers and are subject to certain conditions and restrictions. Student Advantage may add or withdraw benefits at any time.

More Expert Advice

from

THE PRINCETON REVIEW

Find the right school • Get in • Get help paying for it

**CRACKING THE SAT & PSAT
2000 EDITION**
0-375-75403-2 • $18.00

**CRACKING THE SAT & PSAT WITH
SAMPLE TESTS ON CD-ROM
2000 EDITION**
0-375-75404-0 • $29.95
Mac and Windows compatible

**THE SCHOLARSHIP ADVISOR
1999 EDITION**
0-375-75207-2 • $23.00

SAT MATH WORKOUT
0-679-75363-X • $15.00

SAT VERBAL WORKOUT
0-679-75362-1 • $16.00

**CRACKING THE ACT
1999-2000 EDITION**
0-375-75282-X • $18.00

**CRACKING THE ACT WITH
SAMPLE TESTS ON CD-ROM
1999-2000 EDITION**
0-375-75281-1 • $29.95 Mac and
Windows compatible

**DOLLARS & SENSE FOR COLLEGE
STUDENTS**
How Not to Run Out of Money by
Midterms
0-375-75206-4 • $10.95

**STUDENT ADVANTAGE GUIDE TO
COLLEGE ADMISSIONS**
Unique Strategies for Getting into the
College of Your Choice
0-679-74590-4 • $12.00

**PAYING FOR COLLEGE WITHOUT
GOING BROKE
1999 EDITION**
Insider Strategies to Maximize Financial
Aid and Minimize College Costs
0-375-75211-0 • $18.00

**BEST 331 COLLEGES
2000 EDITION**
The Buyer's Guide to College
0-375-75411-3 • $20.00

**THE COMPLETE BOOK OF COLLEGES
1999 EDITION**
0-375-75199-8 • $26.95

**THE GUIDE TO PERFORMING ARTS
PROGRAMS**
Profiles of Over 600 Colleges, High
Schools and Summer Programs
0-375-75095-9 • $24.95

With Free *Apply!* Software

We also have books to help you score high on the SAT II, AP, and CLEP exams:

Cracking the AP Biology Exam 1999-2000 Edition
0-375-75286-2 • $16.00

Cracking the AP Calculus Exam AB & BC 1999-2000 Edition
0-375-75285-4 • $17.00

Cracking the AP Chemistry Exam 1999-2000 Edition
0-375-75287-0 • $17.00

Cracking the AP English Literature Exam 1999-2000 Edition
0-375-75283-8 • $16.00

Cracking the AP U.S. Government and Politics Exam 1999-2000 Edition
0-375-75288-9 • $16.00

Cracking the AP U.S. History Exam 1999-2000 Edition
0-375-75284-6 • $16.00

Cracking the AP Physics 1999-2000 Edition
0-375-75289-7 • $16.00

Cracking the AP European History 1999-2000 Edition
0-375-75290-0 • $16.00

Cracking the CLEP 1999 Edition
0-375-75212-9 • $20.00

Cracking the SAT II: Biology Subject Test 1999-2000 Edition
0-375-75297-8 • $17.00

Cracking the SAT II: Chemistry Subject Test 1999-2000 Edition
0-375-75298-6 • $17.00

Cracking the SAT II: English Subject Test 1999-2000 Edition
0-375-75295-1 • $17.00

Cracking the SAT II: French Subject Test 1999-2000 Edition
0-375-75299-4 • $17.00

Cracking the SAT II: History Subject Test 1999-2000 Edition
0-375-75300-1 • $17.00

Cracking the SAT II: Math Subject Test 1999-2000 Edition
0-375-75296-X • $17.00

Cracking the SAT II: Physics Subject Test 1999-2000 Edition
0-375-75302-8 • $17.00

Cracking the SAT II: Spanish Subject Test 1999-2000 Edition
0-375-75301-X • $17.00

 Visit Your Local Bookstore or Order Direct by Calling 1-800-733-3000

FIND US...

International

Hong Kong
4/F Sun Hung Kai Centre
30 Harbour Road, Wan Chai,
Hong Kong
Tel: (011)85-2-517-3016

Japan
Fuji Building 40, 15-14
Sakuragaokacho, Shibuya Ku,
Tokyo 150, Japan
Tel: (011)81-3-3463-1343

Korea
Tae Young Bldg, 944-24,
Daechi- Dong, Kangnam-Ku
The Princeton Review- ANC
Seoul, Korea 135-280,
South Korea
Tel: (011)82-2-554-7763

Mexico City
PR Mex S De RL De Cv
Guanajuato 228 Col. Roma
06700 Mexico D.F., Mexico
Tel: 525-564-9468

Montreal
666 Sherbrooke St.
West, Suite 202
Montreal, QC H3A 1E7 Canada
Tel: (514) 499-0870

Pakistan
1 Bawa Park - 90 Upper Mall
Lahore, Pakistan
Tel: (011)92-42-571-2315

Spain
Pza. Castilla, 3 - 5º A, 28046
Madrid, Spain
Tel: (011)341-323-4212

Taiwan
155 Chung Hsiao East Road
Section 4 - 4th Floor,
Taipei R.O.C., Taiwan
Tel: (011)886-2-751-1243

Thailand
Building One, 99 Wireless Road
Bangkok, Thailand 10330
Tel: (662) 256-7080

Toronto
1240 Bay Street, Suite 300
Toronto M5R 2A7 Canada
Tel: (800) 495-7737
Tel: (716) 839-4391

Vancouver
4212 University Way NE,
Suite 204
Seattle, WA 98105
Tel: (206) 548-1100

National (U.S.)

We have over 60 offices around the U.S. and run courses in over 400 sites. For courses and locations within the U.S. call **1 (800) 2/Review** and you will be routed to the nearest office.